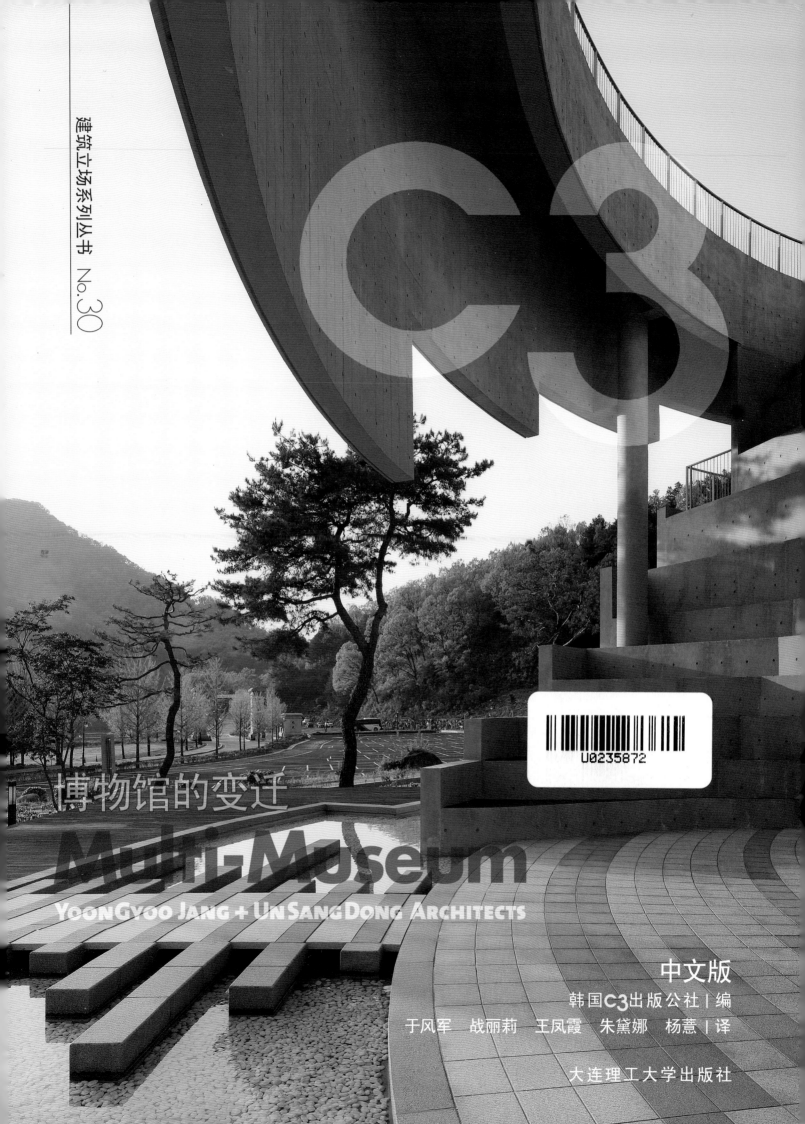

建筑立场系列丛书 No.30

博物馆的变迁

Multi-Museum

YoonGyoo Jang + UnSangDong Architects

中文版

韩国C3出版公社 | 编

于风军　战丽莉　王凤霞　朱黛娜　杨薏 | 译

大连理工大学出版社

博物馆的变迁

004 当代博物馆 _ *Silvio Carta + Marta Gonzàlez Anton*

012 帕里什艺术博物馆 _ *Herzog & de Meuron*

026 卢浮宫朗斯分馆 _ *SANAA*

040 佩洛特自然和科学博物馆 _ *Morphosis Architects*

056 大玛雅文明博物馆 _ *Grupo Arquidecture*

070 宽容与记忆博物馆 _ *Arditti + RDT Arquitectos*

084 荷兰新国家博物馆 _ *Cruz y Ortiz Arquitectos*

098 国立 Machado de Castro 博物馆 _ *Gonçalo Byrne Arquitectos*

116 特里亚纳陶瓷博物馆 _ *AF6 Arquitectos*

YoonGyoo Jang+UnSangDong建筑师事务所

128 社会想象 _ *YoonGyoo Jang*

130 由复合体向社会想象体的转变 _ *YoungBum Reigh*

134 Water Circle博物馆——CheongShim水文化中心

152 文化街——松洞文化与福利中心

166 运动想象——2012韩国YeoSu世博会现代汽车展馆

178 Rooftecture——能源与绿色住宅

Multi-Museum

004 *The Contemporary Museum* _ Silvio Carta + Marta Gonzàlez Anton

012 Parrish Art Museum _ Herzog & de Meuron

026 Louvre Lens Museum _ SANAA

040 Perot Museum of Nature and Science _ Morphosis Architects

056 Great Museum of the Maya Civilization _ Grupo Arquidecture

070 Museum of Memory and Tolerance _ Arditti + RDT Arquitectos

084 The New Rijksmuseum _ Cruz y Ortiz Arquitectos

098 Machado de Castro National Museum _ Gonçalo Byrne Arquitectos

116 Triana Ceramic Museum _ AF6 Arquitectos

YoonGyoo Jang + UnSangDong Architects

128 *Social Imagination* _ YoonGyoo Jang

130 *Transformation of Compound Body to Social Imaginative Body* _ YoungBum Reigh

134 Water Circle – CheongShim Water Culture Center

152 Culture Street – SeongDong Cultural & Welfare Center

166 Motion Imagination – 2012 YeoSu EXPO Hyundai Motor Group Pavilion

178 Rooftecture – Energy + Green Home

博物馆的变迁
Multi-Museum

对空间行为进行的近期研究(Bill Hillier等人)已经证明人们在当代博物馆内的所作所为受到了建筑的具体形式(而非其展品)的极大影响。这一权威的发现将有助于在长期的争论中将文化建筑置于一个明确的位置。究竟文化建筑是一个"容器"还是一个"被包容"的物体?因此,同时也应考虑到当代博物馆的本质,以及其内部层次布局。本期讨论的项目将跨越很长的时间跨度:在过去的几十年里,我们已经见证了博物馆定义所发生的巨大变化,即作为一座公共建筑及其城市中的一员,更广泛的说,作为文化和社会中的一员。其典型的案例包括从博物馆主要的吸引力和特色根源是其展品的"尊贵的容器"(参考位于伦敦的约翰·索恩爵士住宅),到那些极具唤醒力和刺激性以至于吸引每个人目光的项目(如盖里的古根海姆博物馆),再到其本身便是展品一部分的博物馆。随着时间的流逝,建筑师及其委托人对博物馆的传统概念发动了挑战,其概念与一些标准型的假设对博物馆的可能性地点及其社会角色起到推动作用。在建筑现代史中,少数的博物馆(如纽约的古根海姆博物馆及其盘旋的展览空间,或者是Lina Bo Bardi的Sān Paulo艺术博物馆及其非凡的公共空间)已经表明了一个明确的立场,即关于当代博物馆的本质,以及其所代表的角色的争论。然而,这个问题仍然存在:今天,当代博物馆应该是什么形象?博物馆怎样将当代社会的文化、娱乐以及知识需求转化为其具有战略性的具体形式?博物馆怎样与城市及其居民相联系?在信息时代,博物馆能为用户提供怎样的体验?此外,博物馆是为每一个人,还是仅仅为一小部分专家而服务?本期所展示的项目为我们的未来提供不同的建筑方法和方案,以给予我们帮助。即使这些方法和方案没有给予明确的回答,至少也给予了这些问题大致的框架范围。

Recent studies of spatial behavior (Bill Hillier et al.) have demonstrated that people's behavior in contemporary museums is significantly more affected by the physical forms of the building than by what it exhibits. This definitive finding contributes to establishing a clear position in the longstanding debate concerning cultural buildings – container or contained? – and hence concerning what a contemporary museum should be, or how its internal hierarchies should be arranged. This discussion has a long history: In the last few decades we have witnessed significant changes in the notion of the museum as a public building and its role for the city and – more broadly – for culture and society. Examples stretch from noble containers in which the main attraction and the source of its distinctive character are the exhibits the museum houses (think of the Sir John Soane House in London) to projects that are so powerfully evocative and breathtaking as to attract all the attentions per se (e.g. Gehry's Guggenheim Bilbao) until the museums themselves become the main piece on exhibit. Over time, architects and their commissioners have challenged traditional conceptions of the museum, and with them, standard assumptions are connected to its possible place in the city and its role in society. In the recent history of architecture, a few museums (e.g. the Guggenheim in New York and its spiral exhibition space, or Lina Bo Bardi's São Paulo Museum of Art and the extraordinary public space it produces) have epitomized a clear position in the debate concerning what the contemporary museum should be and represent. Still, the question remains: what should a contemporary museum look like today? How do museums translate contemporary society's demands for culture, entrainment and knowledge into strategies and physical forms? How should museums engage the city and its inhabitants? What user experiences should museums offer in the information age? Moreover, should the museum be for everybody, or for a small niche of experts? The projects presented in this issue offer a variety of approaches and solutions for the future of our cities which can help – if not definitively answer – at least frame such questions.

帕里什艺术博物馆_Parrish Art Museum / Herzog & de Meuron
卢浮宫朗斯分馆_Louvre Lens Museum / SANAA
佩洛特自然科学博物馆_Perot Museum of Nature and Science / Morphosis Architects
大玛雅文明博物馆_Great Museum of the Maya Civilization / Grupo Arquidecture
宽容与记忆博物馆_Museum of Memory and Tolerance / Arditti + RDT Arquitectos
荷兰新国家博物馆_The New Rijksmuseum / Cruz y Ortiz Arquitectos
国立Machado de Castro博物馆_Machado de Castro National Museum / Gonçalo Byrne Arquitectos
特里亚纳陶瓷博物馆_Triana Ceramic Museum / AF6 Arquitectos
当代博物馆_The Contemporary Museum / Silvio Carta + Marta Gonzàlez Anton

与许多其他建筑类型学一样，博物馆在人类历史上经历了一些变化。博物馆的理念起源于对某一种价值的占有意识，无论这个价值是一种艺术品，还是一段历史，抑或是珍品，都值得对外展出。围绕着这一意愿，人们便会留意对过去的事物的保存、稳定物价的意图（以在此时此地对什么的最重要的规划出一个明确的立场，或者维护一种特殊文化）。

博物馆的理念深深扎根于社会意识中，这是一种个人或是社会团体所具有的、关于艺术品与历史文物的意识。回顾早期的博物馆，人们可能会回想起卡皮托利尼博物馆（建于15世纪的第一座面向公众开放的艺术品收藏馆），或者是约建于16世纪的梵蒂冈博物馆。值得注意的是，早期的博物馆向大多数公众或者是部分公众展示大量的展品，而这些展品是有一些权力人物，如基督教教皇或者是富有的贵族来决定的。

尽管一些18世纪前的案例也值得人们注意，但是博物馆兴起的时间主要还是从启蒙运动时期开始。那么，博物馆的展品以及由宣传所引发的争议是一场激进的改革（被认为启蒙运动的社会已经开始）也就不足为奇了。如狄德罗和达朗贝尔的《百科全书》（1751—1772年），书中展示的当时法国专业技术知识的细节，并且将其受众面扩至绝大部分的公众，博物馆开始将艺术、文物和珍品向公众开放。从这方面来说，人们可能希望面对一种考虑到艺术以及大部分公众的古老方式，并且回想起约翰·伯格于1972年的著作《观看之道》，这一纪录文献解释了启蒙运动之前和期间大部分私人委托绘制的裸体画像较为普遍的原因。裸露是将一个主题（裸体女人）具体化，以使观看者获得较为纯粹的愉悦感。因此，油画的主人，即一个身心健康的人，买下油画，以独自获得这种愉悦感，或者向他的客人展示他的权力。在这种程度上，很明显，艺术便专属于一小部分社会，而不是大部分公众，也从不有意更广泛地被展览出来。

在启蒙运动期间，博物馆面向公众的开放性逐渐扩大，且极具进步性地实行民主化，这是一个跨越式的跃进。另一座当代博物馆的先驱没有将重点放在艺术上，而是在珍品方面。所谓的"珍品陈列室"或者"Wunderkammer"即是一个私人房间，贵族或者商人经常来此光顾，这里展示了各种类型的珍品，它们有可能来自异国的探险途中，抑或是遥远的海岛中，包含文物、工具、献纳品、石制品等等。在早期的艺术博物馆中，主要的理念就是再一次地将引起人们兴趣的若干件物品收藏在一个房间内。

目前为止，这座城市与博物馆之间并没有联系，前者是与个人（而非

As with many other architectural typologies, museums have undergone several changes throughout human history. The idea of the museum originated from an awareness of possessing a piece of a certain value – whether an object of art, history or simply curiosity – worthy of being displayed to others. Around this willingness, one may observe in the preservation of past items and their valorization an intention to formulate a clear stance on what is considered important *hic et nunc* in a specific time, or to affirm a specific culture.

The idea of the museum is deeply rooted in the societal view that private or public bodies may have of art and historical artifacts. Amongst the early museums one may recall the Capitoline Museums – one of the first collections of art open to the public in the 15th Century – or the Vatican Museums from around the 16th Century. It is noteworthy that early museums offered to a wide or a selected public a generous display that powerful persons such as Christian Popes or rich nobles had decided to offer.

Although several examples may be observed from before the 18th century, museums flourished mainly from the Age of Enlightenment onwards. It is no wonder that the exhibition of objects and the debates their promulgation may trigger were a result of the radical change in thinking enlightenment society sought to set in motion. Just as Diderot and d'Alembert's great Encyclopédie (1751~1772) showed in detail the technical expertise of France at the time, expanding that knowledge to a larger public, museums began to open art, historical objects and curiosities to the public. In that regard, one may want to confront a former way of considering art and the larger public, recalling John Berger's 1972 "Way of Seeing". This documentary explains the diffused raison d'etre of most privately commissioned nude portraits in European paintings before and even during the Enlightenment: The nude was an objectification of the subject (the naked woman) for the pure enjoyment of the spectator, in that case the oil painting's owner, a healthy person who could afford the commission of an oil painting for his private enjoyment and to display his power to his guests. To this extent it is clear that art was addressed to a niche of society and was not meant for a larger public, nor was it intended to be widely displayed.

The great leap forward which the Enlightenment gradually introduced was the increased openness of the museum to the public and the progressive democratization of its contents. Another ancestor of the contemporary museum emphasized not the arts, but curiosities. The so-called *cabinet* of curiosities or *Wunderkammer* was a private room, frequented by aristocrats or merchants, that displayed various types of unusual items. These may have come from exotic expeditions or distant lands, and encompassed relics, tools, votive objects, stones and so on. As with early art museums, the main idea was again to collect in one room several objects

约翰·索恩爵士博物馆, 伦敦
Sir John Soane's Museum in London

康塞巴托里宫殿美术馆, 是卡匹多利尼博物馆三座主要博物馆之一
the Palazzo dei Conservatori, one of the three main buildings of the Capitoline Museums

一个地点和城市）有着深深的渊源。然而, 博物馆对于城市的重要性却在与日俱增, 在20世纪下半叶, 尤其是受到二战影响的城市重建期, 博物馆开始具有了和现在相同的特点。这些博物馆各式各样, 其设计目标和意图也不尽相同, 我们将会提供几个同时期的案例, 以为本期所展示的项目提供一个有用的参考模板。

博物馆作为城市地标

在一座城市中, 博物馆已成为重要的建筑。在一些情况中, 它们已经变得如此受欢迎, 甚至超越了其所处的环境的市界和国界, 而成为全球认可的存在物: 弗兰克·盖里的位于毕尔巴鄂的古根海姆博物馆最初是作为一座可提供大型投资, 以对城市原有的工业区进行重建的建筑而存在, 而最后它却成为了毕尔巴鄂的旗舰。卢浮宫及其玻璃金字塔（贝聿铭, 1988年）成为巴黎最主要的景点之一。位于马德里的埃尔普拉多博物馆, 抑或是圣彼得堡的冬宫博物馆都是欧洲最大的博物馆之一。

博物馆作为私有的荣耀

与公共博物馆（设想将展品面向广泛的公众开放）不同的是, 一些博物馆打算设置某种程度的私密性, 以远离大部分公众的视线或大多数媒体的注意。这些私人画廊仍然信奉一种资产阶级的理想式艺术, 完全依靠个人对其私有艺术品的喜好而定, 并将它们与其专门邀请的客人共享。如此私人的展览空间性质范围较广, 从历史名城中心的古老宫殿, 如弗朗克斯·皮诺特的威尼斯葛拉西宫, 到城堡和名宅, 再到为特别展品和收藏家而设计的私人博物馆。少数的几个典范则有理查德·迈耶设计的弗里德·布尔达收藏馆, 位于德国巴登巴登, 以及赫尔佐格&德梅隆设计的, 位于慕尼黑的私人美术馆, 又或者是Dakis Joannou的Guilty博物馆, 其本身就是一件艺术品。这些私人博物馆都是经过精心设计的, 且以业主的理念和视野为基础, 以反映收藏品和收藏者之间的亲密关系。

博物馆作为权威的机构

其他的博物馆则以权威的专业机构形式出现, 并且致力于展现一个特定的主题。这些成为这个主题的一个主要参考点, 并且用以保护和保存某些特殊的知识。此外, 一些博物馆成为科研院所, 并且将其所负责的信息领域进行传播。历史、考古或者自然科学博物馆都是这一类型的

worthy of spectator interest.

Up to this point the museum and the city shared no relation, with the former being deeply associated with one person, rather than with a place or a city. However, the importance of the museum to the city grew gradually but extensively throughout the second half of the twentieth century, at which time – especially during the reconstruction of several cities affected by the World Wars – it began to acquire some of the features by which we recognize museums today. These are quite varied and can be designed with a variety of aims and purposes. We will provide here a few coordinates in an attempt to construct a sufficiently large frame of reference for this issue's presented projects.

Museum as City Landmark

Museums have become important buildings for the city. In some cases they have become so popular as to transcend the municipal or even national boundaries of their contexts and to become globally recognized presences: Frank Gehry's Guggenheim in Bilbao originated as an architectural provision of a huge investment intended to renovate the city's former industrial area, eventually became Bilbao's flagship. Musée du Louvre and its Pyramid (I. M. Pei, 1988) are among the main attractions of Paris and indeed of France. Museums such as el Prado in Madrid or State Hermitage in Saint Petersburg are among Europe's largest museums.

Museum as Private Pride

Unlike public museums – that is, those conceived to open their content to a wider public – some have been intended to be somewhat élitaire, to escape the view of the public at large and the attention of the general media. These private galleries still espouse a sort of bourgeois ideal of art, hinging on one's private enjoyment of one's own art pieces and on sharing them with a selected audience. Such private exhibition spaces may range from old palaces in historical city centers, such as François Pinault's Palazzo Grassi in Venice, to castles and stately homes, to private museums designed ad hoc for a specific collection or collector. A few examples are Richard Meier's Museum Frieder Burda in Baden-Baden, Germany; Munich's Sammlung Goetz by Herzog & de Meuron; or the extreme case of Dakis Joannou's Guilty, which is an artwork in its own right. Such private museums are carefully designed around the ideas and vision of their owners, reflecting an intimate liaison between collection and collector.

Museum as Authoritative Institution

Other museums take the form of authoritative institutions dedicated to a specific subject. These become a main reference for the subject and serve to conserve and preserve certain specific knowledge. Moreover, some become research institutes and diffuse the information they are responsible for. Historical, archaeological or

埃尔普拉多博物馆,是世界上参观次数最多的场地,被认为是最宏伟的博物馆之一
El Prado Museum in Madrid, one of the most visited sites in the world, is considered to be among the greatest museums

俄罗斯圣彼得堡的冬宫博物馆,是世界上最大、最古老的博物馆之一
The State Hermitage in Saint Petersburg, Russia, one of the largest and oldest museums in the world

较为明显的典范。

博物馆作为文化推进器

其他博物馆则是作为文化推进器而存在着。在这些博物馆中,一些最初的趋势以及文化运动可能在全球引起共鸣。一个较为明显的例子便是纽约城的现代艺术博物馆,它起源于一种"国际风格:1922年后的建筑"(Johnson、Barr和Hitchhock,1932年)以及解构主义建筑(Johnson和Wigley,1988年)展览中,两者都对整个世界的建筑产生了重要的影响。这类图书馆不再扮演文化卫道者的传统角色,而是更积极地迈出了一步,揭开了更加新奇且进步的新篇章。

就地存在的博物馆

特别是在重大考古发现或是雕塑作品领域,博物馆可以以外光画展览空间的形式出现。位于开放式场地的地块,如考古场地,或者是一些地理特点(规模以及位置等)使其无法围合起来的场地,都以博物馆的名义框在一处区域内,这一地块范围从赫库兰尼姆和庞培的考古区域,到带有室外雕塑的大型公园,再到整个旧城中心。

博物馆作为一处有魔力的地方

艺术品与其独特的自然环境之间的强烈纽带还可以催生一种以场地为基础的博物馆,其建筑特点是试图与室外空间建立有趣的联系。因此,艺术不再仅仅是含括在博物馆其中,或者是作为展品来展出,而是成为整个博物馆体验的一部分。位于哥本哈根北部的路易斯安那现代艺术博物馆以及位于里斯本的Gulbenkian博物馆便是带有这一设计目标的典型案例。

博物馆本身便是一处创新设计

某些博物馆记录了展览空间的历史,即便没有展示整座建筑的全景,博物馆也详细地展示了创新的理念,使参观者得到了丰富的体验。弗兰克·劳埃德·赖特的纽约古根海姆博物馆使这座1959年建成的博物馆的理念发生了变革性的变化。其连续的展览空间在建筑物的外边缘上不断变化,沿着盘旋而上的坡道从地面上升至顶部,打破了走廊(或者是类似于走廊的空间)与展览空间之间的常规不同点。赖特的古根海姆博物馆的空间如此的引人注目,以至于它引起了人们的担忧,即这座建筑

natural science museums are clear examples of this category.

Museum as Cultural Propeller

Other museums work as cultural promoters, and from them originate trends or cultural movements may resonate globally. One clear example is the Museum of Modern Art in New York City, from which originated the "International Style: Architecture Since 1922" (Johnson, Barr, and Hitchcock, 1932) and Deconstructivist Architecture (Johnson and Wigley, 1988) exhibitions, both of which significantly influenced the entire world of architecture. Instead of the traditional role of careful guardian of culture, in such cases the museums take positive steps, proposing novelty and advancing possible new scenarios.

Just-there Museums

Especially in the field of significant archaeological findings or sculptural artworks, the museum can assume the form of a plein-air exhibition space. Pieces already in such open fields as archaeological sites, or whose physical characteristics (dimensions, location, etc.) do not lend themselves to enclosure are simply framed within an area which takes the name of a museum. These range from the archaeological areas of Herculaneum and Pompeii to large parks with open-air sculptures, to entire old city centers.

Museum as Magic Place

Site-based museums can also emerge from a strong marriage between artworks and unique natural surroundings. Architectural features here try to establish interesting relationships with the outdoor spaces. The art is hence no longer contained or merely exhibited within, but becomes part of an overall experience. The Louisiana Museum of Modern Art in north Copenhagen and the Gulbenkian Museum in Lisbon are clear cases of such a design intention.

Museum as Innovation Per Se

Certain museums have marked the history of exhibition spaces, if not of the entire architectural panorama, by elaborating an utterly innovative concept and visitor experience. Frank Lloyd Wright's Guggenheim in New York revolutionized the entire idea of the museum in 1959. Its continuous exhibition space evolving on the outer edges of the building along a spiral ramp from the ground to the top broke the usual distinction between corridors (or spaces as such) and proper exhibition spaces. The spatiality of Wright's Guggenheim proved so compelling that it has raised concerns that the architecture *per se* may become more important and relevant than the artworks it houses. New York's Guggenheim represents those museums whose primary focus is the building itself and the visitor experience.

弗朗克斯·皮诺特的威尼斯葛拉西宫，由Giorgio Massari设计
François Pinault's Palazzo Grassi in Venice designed by Giorgio Massari, 1748~1772

赫尔佐格&德梅隆设计的、位于慕尼黑的私人博物馆，1992年
Sammlung Goetz in Munich, Germany by Herzog & De Meuron, 1992

本身可能会变得比其展出的艺术品更加重要，更具有意义。在那些最初的焦点是建筑本身以及参观者体验的博物馆中，纽约古根海姆博物馆则是其中的代表性作品。

博物馆作为一个品牌

对于博物馆来说，能够帮助配置其现代功能转换的另一方面是所有博物馆都会拥有的一系列辅助操作。除了展览空间以及相关的活动之外，博物馆还设有餐馆、咖啡室、酒吧、书店、图书馆、学习室以及会议室，并且还设有出售一系列饰品的售货区，出售范围从铅笔刀到T恤，所有这些商品都致力于创造一个全新的品牌。品牌化的博物馆成为以其品牌来吸引更多游客的一座博物馆。

博物馆作为城市社会的一个构件

一些博物馆的主要目的似乎是与城市及其居民建立友好的关系。赫尔佐格&德梅隆设计的位于伦敦的涡轮大厅为伦敦人提供一处特殊的公共空间，而这处空间不仅仅与展览室直接相连。类似的，Lina Bo Bardi的São Paulo艺术博物馆（1947年）则在与街道齐平的一侧建造了一处超凡的公共空间，吸引人们进入其范围内。越来越多的公共建筑试图与城市建立更强烈的关系，并不仅仅旨在提供公共空间，而是要模糊展览空间和其周围的休闲区域之间的界限（以至于博物馆影响范围的物理界限变得模糊）。

在过去的几十年里，一种具有进步性的混合型博物馆出现了。从一座带有清晰的设计目标的单体建筑，博物馆逐渐转换为一座混合型建筑，并且优于本文所展现的几个类型。同样的，当代博物馆在其功能、空间规划以及其所提出的展览规程方面，是属于部分创新的。然而，博物馆将其自身看做是一处充满想象力的空间，同时与城市的其他地区建立全新且重要的联系，并且扮演着当地文化特殊部分的管理者的角色，十分有影响力。而此处所规划的焦点则是如今许多博物馆试图以不同的方式和方法，来实现所有的规划。因此，将任何一座现代博物馆项目进行单一的分类是十分困难的。本期所展示的项目均是令人信服的案例，没有人能将上述的案例单一地划分到主要的类别中。它们作为过去博物馆的一个缩影，展示了一种思考类别的方式。

例如，卢浮宫博物馆朗斯分馆，由SANAA建筑事务所设计，致力于其所在地区的经济复苏，曾是一处繁华的矿业和重工业地区，这一经典

Museum as Brand

Another aspect that may help configure the contemporary transformation of the museum is the array of side activities all museums appear to have. In addition to exhibition spaces and related activities, museums now house restaurants, cafeterias, bars, bookshops, libraries, study rooms, and conference rooms, and offer for sale a long list of trinkets, from pencils to t-shirts, all of which contributes to creating a museum brand. The branded museum becomes an object of desire that attracts more visitors by the allure of its name.

Museum as Urban Social Component

The main aim of some museums appears to be to engage with the city and its inhabitants. The Turbine Hall of Herzog & de Meuron's Tate Modern in London(2000) offers Londoners a special public space not directly related to the exhibition rooms. Similarly, Lina Bo Bardi's São Paulo Museum of Art(1947) produces an extraordinary public space at the street level, magnetizing people to its footprint. Increasingly, public buildings try to engage the city more widely, not only offering free public spaces, but blurring the boundaries between exhibition spaces and leisure areas in their immediate surroundings (so that the physical limit of the museum's influence becomes undefined).

There has emerged a progressive hybridization of all these "types" of museums over the last few decades. From a single building with a clear intent, museums have gradually turned into mixed-purpose settlements, transcending the few categories here presented. By the same token, contemporary museums are in part quite innovative in their programs and spatial organization, as well as in their proposed exhibition strategies, yet they present themselves as fancy places, while triggering new and important relationships with the rest of the city and serving as influential keepers of specific pieces of local culture. The lens here proposed is that many museums nowadays try to serve all these purposes at once, although in different ways and measures. As a consequence, it is quite difficult to categorize any of the contemporary museum projects univocally. The presented projects in this issue provide convincing examples of this reading. None can be univocally defined under any of the main categories above mentioned. They indeed present a way of thinking of this typology as a cross-section of the museums of the past.

Louvre-Lens Museum, for instance, designed by SANAA, was designed to contribute to the economic recovery of its region, once a prosperous mining and heavy industry area. Pursuing an effect similar to that of the Guggenheim on the Bilbao economy – the

理查德·迈耶设计的弗里德·布尔达收藏馆,位于德国巴登巴登,2004年
Richard Meier's Frieder Burda's Museum in Baden-Baden, Germany, 2004

Dakis Joannou的私人Guilty画廊,由Jeff Koons设计,经过精心设计,以业主的理念和视野为基础
Dakis Joannou's private gallery Guilty by Jeff Koons, carefully designed around the ideas and vision of the owners

案例追求类似于毕尔巴鄂古根海姆博物馆的效果,卢浮宫和朗斯(临近英吉利海峡、位于法国北部的城市)当局规划建造一个博物馆园区。博物馆的低层体量与周围景观相结合,其玻璃立面成为室内与室外之间的透明过滤器,而抛光的铝板则反射出博物馆前面的公园的景象。新建筑之间的边界十分模糊:博物馆和城市建立了一种直接的对话。相对于传统的博物馆来说,该建筑占据了一处奇怪的区域,SANAA建筑事务所没有对空间进行围合,而是追求其连续的开放性,或是面向周围景观开放,或是建造一处面积为3000m²、长120m且没有单层内壁的主展览空间。

阿姆斯特丹的新国家博物馆采取类似的具有里程碑意义的方法,旨在宣告其对某一特定艺术氛围的优势地位。经过全面的整修与扩建(十年来已经将博物馆在荷兰所发挥的卓越作用彻底地改变)之后,博物馆重新面向公众开放。这座具有象征意义的博物馆是由荷兰建筑师Pierre Cuypers于1885年设计的,用以收藏国家艺术展品,即艺术高峰期——荷兰黄金时代(17世纪)的画作。这一时期是荷兰荣耀和海上霸权时代的代表。此外,具有纪念意义的国家博物馆还在相当密集的城市肌理中发挥着重要作用。一条公共通道贯穿其中心轴线,两侧建有两座塔以凸显博物馆的地标作用,以及历史城市中心、博物馆广场的文化中心以及在南部与其相邻的通道(每日有成千辆自行车通过)之间的连接作用。

Cruz y Ortiz建筑事务所赢得了改造这座建筑的设计竞赛,即重新恢复其昔日的荣耀,设置新的设施,以满足现代需求,并且还为博物馆增建了几个建筑体量,以容纳相关服务设施。建筑师以明确的姿态来唤起博物馆的城市特点,加以突出。Cruz y Ortiz建筑事务所移走20世纪50—60年代建造的中间楼层,以及降低地面层,以在两座现存的庭院之上建造一个面积为2330m²的大型入口中庭。而在通道下方连接两座庭院的大礼堂,则成为博物馆的真正核心区域:一处大型空间恢复了原始建筑的光度,容纳了主要的服务区(入口、信息台、衣帽间、自助餐厅、餐馆以及书店),还另外设有一个大型市民休息处,人们无需门票便可以来此处休息。

位于达拉斯的佩洛特自然科学博物馆由Morphosis建筑事务所设计,也试图在城市文化全景中发挥重要的作用,使游客能够对自然和科学全景产生实践性和互动性的理解。整个场地沿着城市中心边界的一条最繁忙的马路而建,极具战略意义,周围都是架高的景观。博物馆的

paradigmatic example – the Louvre and the local authorities of Lens, a city in northern France close to the English Channel, proposed a campus of the museum for the city. The museum's low volume integrates with the surrounding landscape; its glazed facade becomes a transparent filter between the exterior and the interior, while the polished aluminum reflects the park in front. The boundaries of the new buildings are subtly blended: Museum and city establish a direct dialogue. The architecture occupies an odd position with regard to traditional museums: Instead of enclosing the space, SANAA pursues its continual opening, either by its attitude towards the surrounding landscape, or its innovative 3,000m² and 120m-long exhibition main space, which has no single inner wall.

Taking a similar landmark approach, The New Rijksmuseum in Amsterdam aims to state its own mastery of a certain artistic sphere. It has recently been reopened to the public after a thorough restoration and extension that over 10 years have significantly transformed the museum par excellence of the Netherlands. The emblematic museum was designed by Dutch architect Pierre Cuypers in 1885 to house the national collection of art, the artistic peak of which – the paintings of the Dutch Golden Age (17th century) – represented the glory and maritime hegemony of the country at that time. Furthermore, the monumental Rijksmuseum has always played an important role in a rather dense urban fabric. A public passage running along its central axis and flanked by two towers highlights the museum as a landmark and as a main connection between the historical city center, the Museumplein cultural hub and the southern neighborhoods – a passage used daily by thousands of cyclists.

Cruz y Ortiz Arquitectos won the competition to transform the building by retrieving its original glory, adapting the facilities to current needs and adding several volumes for museum-related services. Through clear gestures, the architects have invoked and enhanced the urban character of the museum. Cruz y Ortiz have created a large 2,330m² entrance atrium on the two existing courtyards by removing the intermediate levels created in the 50's and 60's and lowering the level of their ground floor. This great hall, which connects the two courtyards below the passage, becomes the true core of the museum: a large space that recovers the luminosity of the original building, houses its main services (entrance, information desk, wardrobe, cafeteria, restaurant and bookshop), and comprises a great urban lounge fully accessible without a museum admission ticket.

Perot Museum of Nature and Science in Dallas by Morphosis Architects also strives to play an important role in the cultural panorama of the city, engaging visitors with a practical and interactive

庞培考古区域,"就地博物馆"的典型案例之一,是一处开放的大型历史展览区域。
Pompeii, one of the examples of "Just-there Museums" being an large open historic exhibition area

主体量浮在基座上方,如同一个封闭的立方体。做为在城市的公共氛围内维持博物馆重要性的一种方法,其整体展览路线将主入口广场(一处位于达拉斯的游客聚集和举办活动的区域和室外活动空间)和博物馆的最高层(一座能看见整座城市中心视野的玻璃阳台)连接起来。如同建筑师所强调的,"公共空间是博物馆必不可少的一部分,就像博物馆是城市必不可少的一部分一样"。

国立Machado de Castro博物馆是与古代重建联系的另一次尝试,这座博物馆位于葡萄牙的科英布拉,由Gonçalo Byrne建筑事务所设计,这座混合型建筑位于科英布拉主教原来的宫殿原址,如今却可以举办各种与博物馆相关的活动,如城市考古场址的保护、博物馆技术、展览、周围所有纪念性建筑的特点的维护。博物馆试图建立一个持久的联系,不仅仅与建筑所处场地的过去相联系,而且还与城市上层部分的整体历史以及知名的科英布拉大学相联系。科英布拉大学是欧洲最古老的大学之一,建于1290年。因此,博物馆通过相互交织的联系以及超越了城市地理界限的关系,成为整座城市至关重要的一部分。

位于墨西哥梅里达的大玛雅文明博物馆同样可以展示这一设计意图,即与城市的社会空间发生联系,这是一个旨在丰富某一文化的考古学、人类学以及民族遗产的项目。这座博物馆由Grupo建筑事务所设计,其所展示的一系列体量的理念基础看起来是毫无关联的,但是一旦你进入这座开放的、被抬高的广告广场,并且走进这座建筑综合体时,各个元素便会依据玛雅理念展示其重要性。根据宇宙观念,其他世界,如被洪水冲毁的世界,便先于玛雅人存在,其观念是宇宙是由四名护卫者所支撑起来的,他们位于四个方位点。玛雅世界的中心点是一棵木棉树,其枝桠向上延伸至天堂,而其根部则伸向地狱。建筑师将这一理念进行了直接的解读,从而设计出了这座新博物馆的体量:一个圆柱形体量,覆有绿色金属枝条,依附在一个正交的体量上面,而这个正交体量被设置成一个整体基座。新博物馆的规模和物质形态将其本身与周围环境清晰的区分开来。

相比之下,位于塞维利亚(西班牙)的特里亚纳陶瓷博物馆的准确位置则与其材料和展出的考古展品有着强烈的联系,并且暗示了这些博物馆是有魔力地方。这个项目由AF6建筑事务所设计,将一座前陶瓷厂修复成一座博物馆。自古便与特里亚纳相邻的区域,即博物馆所在地,与

understanding of natural and scientific phenomena. The whole plot, strategically located along one of the most heavily trafficked roads bounding the city center, is fully colonized by an elevated roofscape. Above this plinth floats the main volume of the museum, conceived as a closed cube. As a way of asserting the importance of the museum within the city's public sphere, the overall exhibition route is designed to connect the main entry square – a gathering and event area for visitors and an outdoor public space for the city of Dallas – to the museum's uppermost level: a glazed balcony featuring a great view over the city center. As the architects emphasize, "the public is as integral to the museum as the museum is to the city".

Another attempt to reconstruct connections with an ancient time is embodied in the Machado de Castro National Museum, Coimbra, Portugal, by Gonçalo Byrne Arquitectos. Situated in the former palace of the bishop of Coimbra, the compound today houses various museum-related activities, such as preservation of the city's ancient archaeological sites, museography, and exhibitions, and the valorisation of the architectural features of all the monumental buildings surrounding it. The museum tries to establish a long-lasting relationship not only with the past of the area in which the building sits, but also with the entire history of the upper part of the city and the renowned University of Coimbra, one of Europe's oldest, established in 1290. Thus the museum becomes a vital part of the entire city by interlacing connections and references that go far beyond its physical boundaries.

A similar intention to engage the urban social space of the city can be found in the Great Museum of the Maya Civilization in Merida (Mexico), a project that aims to dignify the rich archaeological, anthropological and ethnographic heritage of that culture. The museum, designed by Grupo Arquidecture, displays a set of volumes whose conceptual underpinnings are seemingly disconnected. Once you enter the open elevated public plaza and walk inside the complex, the various elements unveil their significance according to the Mayan conception of the world. According to that cosmic conception, other worlds, destroyed by the flood, had existed before their own, with the cosmos supported by four guardians located at the four cardinal points. At the centre of the Mayan world was a ceiba tree whose branches rose to heaven and whose roots penetrated the underworld. Enacting a direct translation of that conception, the architects have designed the new museum's volumes: A cylindrical volume clad in green metal strips rests on orthogonal volumes arranged as an overall plinth. The scale and materialization of the new museum clearly distinguishes its presence from its surroundings.

The precise location of Triana Ceramic Museum in Seville (Spain), by contrast, places it in close relation with the material and archaeological findings it exhibits, and creates the suggestion that

位于丹麦Humlebekde 路易斯安那现代艺术博物馆,其本身成为整个博物馆体验的一部分
Louisiana Museum of Modern Art in Humlebæk, Denmark becoming part of an overall experience itself

赫尔佐格&德梅隆设计的位于伦敦的泰特现代美术馆,2000年,为伦敦市民提供了一处特殊的公共空间
Tate Modern in London by Herzog & de Meuron, 2000, offering Londoners a special public space

整座城市都具有悠久的陶瓷和陶器生产历史,这可以追溯到罗马帝国时代。这个项目重点要突出一层现存的古窑,这里是生产加工,污泥和粘土处理,染色和上釉,以及瓷器烧制过程的展览空间。上次的新空间展出了历史收藏品,而临近的解读中心则为游客提供用于理解特里亚纳周边复杂社会和文化的提示。新空间被现存的立面以及现存临近建筑的高度改变了边界,而且从附近的街道方向来看,它几乎难以被人们所察觉。这个项目没有旨在成为特里亚纳密集的城市景观中的一处视觉参考点,而是当人们进入其中时,作为一处被发现的礼物而存在。

在一些博物馆中,其所倡导的某种价值多过于其作为艺术品收藏容器的功能,而位于墨西哥的记忆与宽容博物馆就值得一提。这座博物馆由Arditti+RDT建筑事务所设计,通过过去种族灭绝的记忆,利用人们的宽容和对话,以在下一代的教育中发现人性的希望。本项目的设计理念便是以这一想法为基础。在这个规划的指导下,主要的展览空间充满了希望,并且被漂浮的儿童纪念馆具体化。博物馆所体现的社会意识已经远超了本地和国界,甚至走向了世界,因为其所描述的罪恶行径已经广泛地产生了共鸣。

同样的,在更广泛的本地层面,位于安普顿南部(美国)的帕里什艺术博物馆成为其所在区域的重要的文化驱动力。当赫尔佐格&德梅隆被要求设计帕里什艺术博物馆的新总部时,它们选择将单一空间的挤压部分建成一处187m长的底层空间,这处空间位于双面坡的屋顶之下。室内均有独立的单元构成。整体的体量让人们想到安普顿南部常被艺术家用作工作室的小型结构,它们的形式较为简单,如谷仓和其他建筑一般。内部布局十分精妙,且非常清晰:几处展览空间和博物馆服务设施沿着脊形交通流线设置,项目仍和其嵌入的周围景观保持着一定的联系。从远处看,其体量如同水平线位置上的一条较低的线条。一旦人们处于悬挑的斜屋顶之下,室外的露台和入口便被钢和木材所渲染,且如景观的自然延伸一般。当地建造方法和简单材料的使用对项目与当地元素之间的联系起到了维持的作用。

those museums are a magical place. The project, designed by AF6 Arquitectos, rehabilitates a former ceramic pottery factory as a museum. Both the traditional neighborhood of Triana, where the museum is located, and the overall city have a long tradition in ceramics and pottery production dating back to the time of the Roman Empire. The project focuses on enhancing the existing ancient kilns on the ground floor where the manufacturing process, the treatment of sludge and clay, the pigmentation and enamel processes and the firing of the ceramic itself are exhibited. The new spaces above display historic collections, while an adjacent interpretation center provides visitors with the key to understanding the intricate social and cultural whole of the Triana neighborhood. The new spaces adjust to the perimeter indicated by the existing facades and the height of existing nearby buildings, while its presence is barely perceptible from the narrow streets nearby. The project is intended not as a visual reference in the dense urban landscape of Triana, but as a gift to be discovered once entered.

Among the museums that act as promoters of certain values more than as containers of art pieces could be mentioned Museum of Memory and Tolerance in Mexico. The concept of the Museum of Memory and Tolerance, designed by Arditti+RDT Arquitectos, is anchored on the idea that through the memory of past genocides, but with tolerance and dialogue, the hope for humanity may be found in the education of new generations. Following that scheme, the main exhibition spaces embrace that hope, materialized in the floating Children's Memorial. The social awareness that the material exhibits goes far beyond the local and national range to the global scale, as it depicts crimes with wide resonance.

Similarly but on a more local level, Parrish Art Museum in Southampton (USA) becomes important driving forces for the culture of the zone in which they are located. When Herzog & de Meuron were asked to design the new headquarters for the Parrish Art Museum, they opted for a low, 187m-long volume resulting from the extrusion of a single height space under a double-pitched roof. The individual units created inside and the overall volume recall the small-scale structures of Southampton often used by artists as studios, in simple forms such as of barns and other farm buildings. The distribution inside is subtly and clearly stated: Several exhibition spaces and museum-related services are located along the circulation spine. The project maintains a connection with the landscape in which it is inserted: From a distance its volume appears as a low line on the horizon. Once one is below the pitched roof cantilevering, the exterior terraces and entrances emerge as a natural extension of that landscape, rendered in a steel and wood structure. The use of local construction methods and simple materials supports the intimate connection between the project and local factors. *Silvio Carta + Marta Gonzàlez Anton*

博物馆的变迁 Multi-Museum

帕里什艺术博物馆
Herzog & de Meuron

帕里什艺术博物馆新馆始于长岛东端的艺术家工作室，建筑师通过提取工作室的比例和采用朝北开有大天窗的简单住宅的剖面来设置这一独栋画廊空间的基本参数。其中两个画廊成为两翼，围绕着一个脊形交通流线而设置，流线与两个门廊相连，形成这一简约建筑的突出部分的基本框架。

这一突出部分的楼层平面图是对这一理想功能布局的直接解读。博物馆中央创造了十个可分割的画廊，它们簇拥在一起，每个画廊的大小和比例在既定的框架网格下通过调整隔墙便很容易地改变。画廊中心的东部设有行政区、存储区、工作坊和展品装载区等博物馆后勤管理机构。西部则设有公共服务区，有休息大厅、商店、咖啡馆，最西端是一个灵活的多功能教育空间。

井然有序的立柱、横梁和桁架定义了整齐划一的建筑主体。建筑材料就地取材，并且使用当地施工方法。对整个建筑形式来说，现浇混凝土外墙就如同书架上长长的书挡，这些规模宏大的外墙底部与一个连续的长椅连在一起，同时可供游客就座休息、欣赏周围景色。横跨整座建筑全长的大型悬臂结构形成了门廊和露台，为户外活动提供了庇护空间。

天窗的朝向——北面直接决定了建筑的布局，博物馆为东西走向，巧合地与场地形成对角线关系，使整座建筑透视图极富变化，进一步强调了此建筑极端而简单的比例。整座建筑匍匐在宽广的草地上，与长岛的自然景观融为一体。

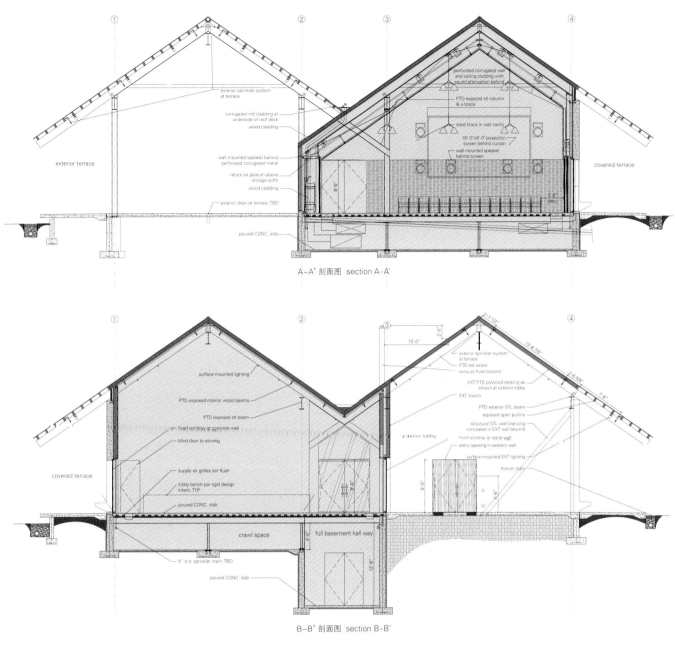

A-A' 剖面图 section A-A'

B-B' 剖面图 section B-B'

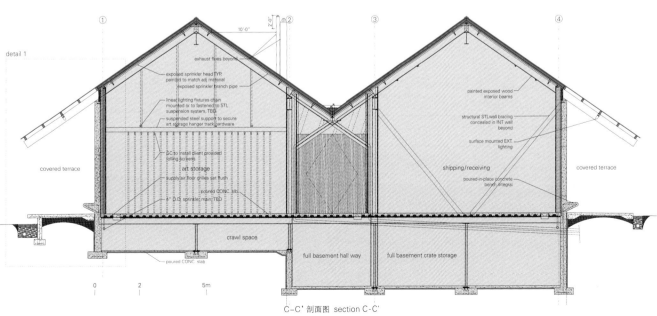

C-C' 剖面图 section C-C'

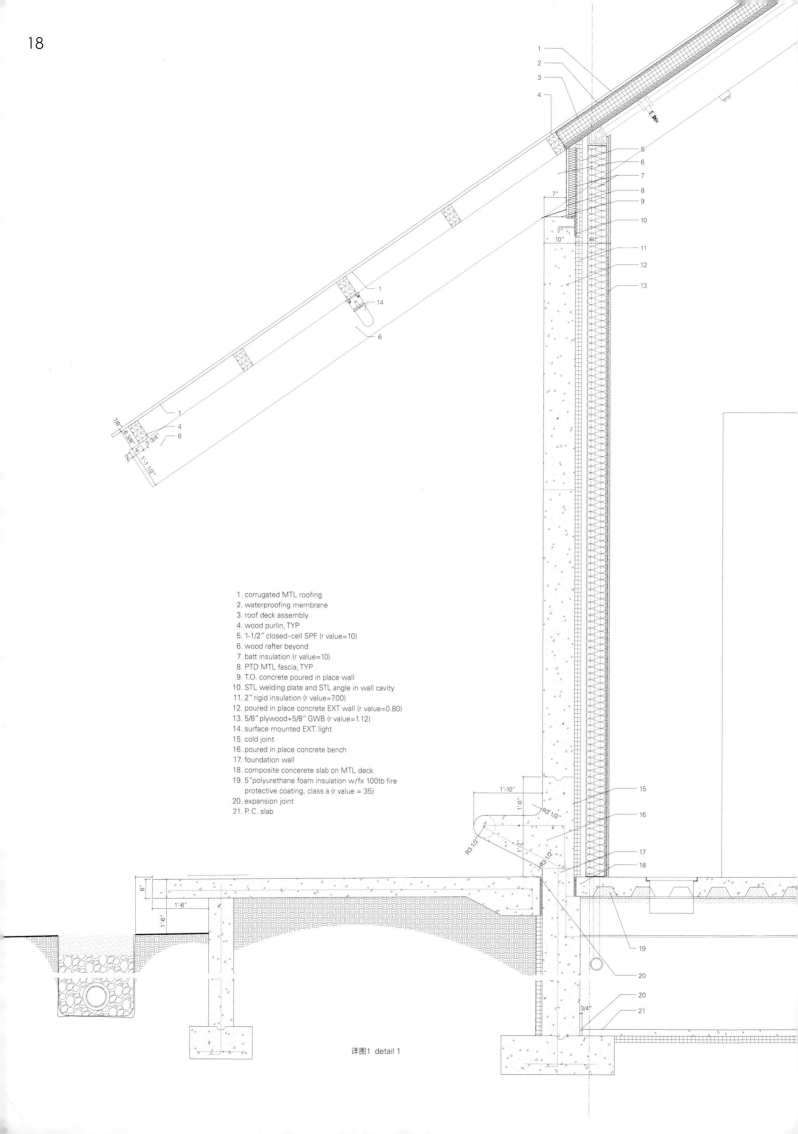

详图1 detail 1

1. corrugated MTL roofing
2. waterproofing membrane
3. roof deck assembly
4. wood purlin, TYP
5. 1-1/2" closed-cell SPF (r value=10)
6. wood rafter beyond
7. batt insulation (r value=10)
8. PTD MTL fascia, TYP
9. T.O. concrete poured in place wall
10. STL welding plate and STL angle in wall cavity
11. 2" rigid insulation (r value=7.00)
12. poured in place concrete EXT wall (r value=0.80)
13. 5/8" plywood+5/8" GWB (r value=1.12)
14. surface mounted EXT. light
15. cold joint
16. poured in place concrete bench
17. foundation wall
18. composite concrete slab on MTL deck
19. 5" polyurethane foam insulation w/fx 100tb fire protective coating, class a (r value = 35)
20. expansion joint
21. P. C. slab

Parrish Art Museum

The starting point for the new Parrish Art Museum is the artist's studio in the east end of Long Island. The architect set the basic parameters for a single gallery space by distilling the studio's proportions and adopting its simple house section with north-facing skylights. Two of these model galleries form wings around a central circulation spine that is then bracketed by two porches to form the basis of a straightforward building's extrusion.

The floor plan of this extrusion is a direct translation of the ideal functional layout. A cluster of ten galleries defines the heart of the museum. The size and proportion of these galleries can be easily adapted by re-arranging partition walls within the given structural grid. To the east of the gallery core is located the back of house functions of administration, storage, workshops and loading dock. To the west of the galleries are housed the public program areas of the lobby, shop, and cafe with a flexible multi-purpose and educational space in the far western end.

An ordered sequence of post, beam and truss defines the unifying backbone of the building. Its materialization is a direct expression of accessible building materials and local construction methods. The exterior walls of in-situ concrete act as long bookends to the overall building form, while the grand scale of these elemental walls is tempered with a continuous bench formed at its base for sitting and viewing the surrounding landscape. Large overhangs running the full length of the building provide shelter for outdoor porches and terraces.

The placement of the building is a direct result of the skylights facing towards the north. This east-west orientation, and its incidental diagonal relationship within the site, generate dramatically changing perspective views of the building and further emphasize the building's extreme yet simple proportions. It lays in an extensive meadow of grasses that refers to the natural landscape of Long Island.

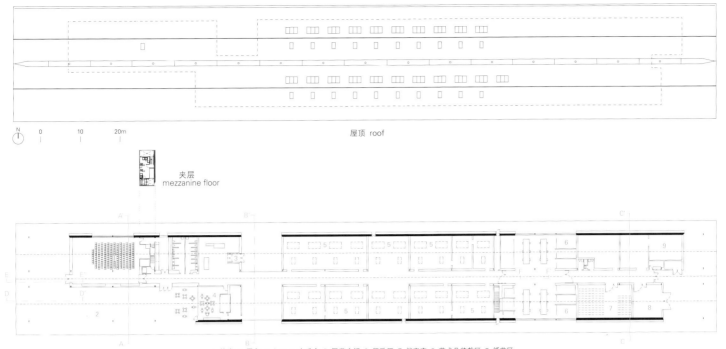

1 礼堂 2 露台 3 入口 4 咖啡室 5 展览空间 6 行政区 7 档案室 8 艺术品装载区 9 纸艺区
1. auditorium 2. terrace 3. entrance 4. cafe 5. exhibition spaces 6. administration 7. archive 8. art loading 9. works on paper

项目名称：Parrish Art Museum
地点：Water Mill, New York, USA
建筑师：Jacques Herzog, Pierre de Meuron, Ascan Mergenthaler (partner in charge)
项目团队：Philip Schmerbeck (associate, project director), Jayne Barlow (associate), Raymond Jr. Gaëtan, Jack Brough, Marta Brandão, Sara Jacinto, Tom Powell, Nils Sanderson, Leo Schneidewind, Camia Young
结构工程师：S.L. Maresca & Associates
机械工程师：Buro Happold
照明设计师：ARUP Lighting
景观建筑师：Reed Hilderbrand Associates Inc
总承包商：Ben Krupinksi Builders
甲方：Parrish Art Museum
功能：visitor services, cafe, multipurpose theater, galleries, administrative offices, archive, storage
用地面积：55,890m² 有效楼层面积：4,673m²
设计时间：2009.3—2010.5 施工时间：2010.7—2012.11
摄影师：©Roland Halbe

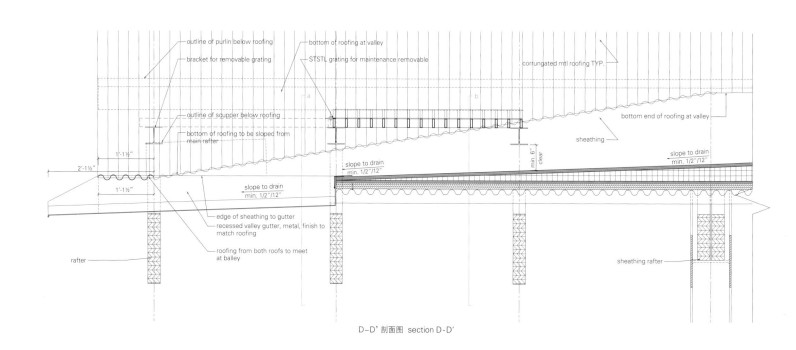

D-D' 剖面图 section D-D'

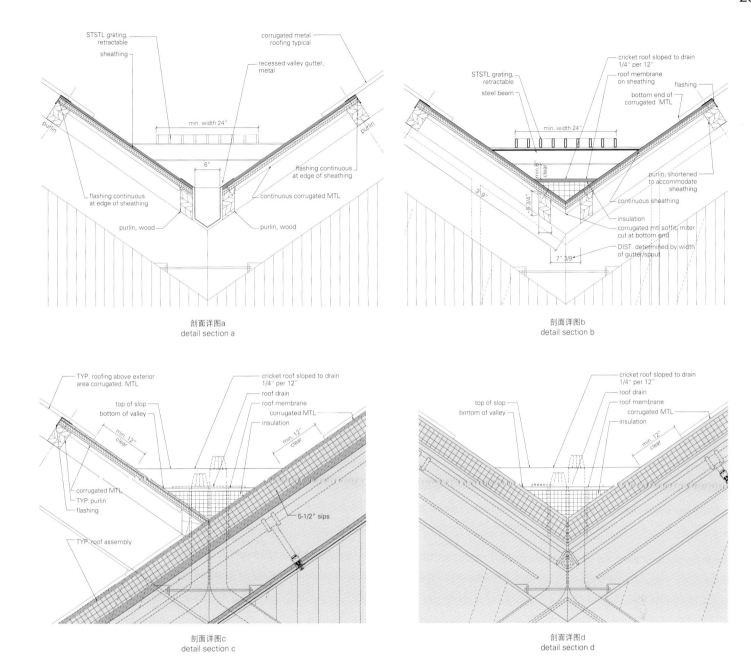

剖面详图a
detail section a

剖面详图b
detail section b

剖面详图c
detail section c

剖面详图d
detail section d

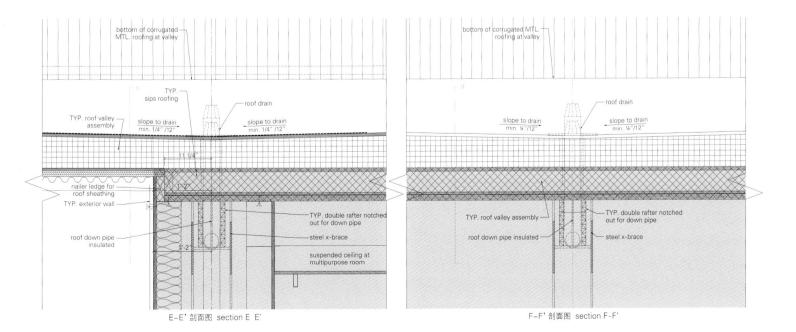

E-E' 剖面图 section E-E'

F-F' 剖面图 section F-F'

卢浮宫朗斯分馆

SANAA

为了保持场地的开放性、减少这一庞大工程的支配地位,设计师把卢浮宫朗斯分馆分解成几处空间。根据地形高程的渐缓变化,通过不同的空间大小和布局,建筑物与场地规模、道路形状和可以唤起人们对当地采矿历史记忆的景观特色达成了平衡。

为了使场地在视觉上和实际上真正开放,位于核心位置的是一座中空的玻璃主体建筑。这个精致的玻璃箱是整个博物馆的入口大厅,是朗斯市一处真正的公共空间。它是透明的,通向整个区域的几个不同方向,穿过它可以到达城市的不同角落。

该项目设计没有采用绝对的直线形状,那样将与场地的微妙特点格格不入;也没有采用天马行空的自由形状,因为从博物馆的内部运作角度来说,这会受到特别大的限制。项目设计采用了几个空间的轻微回折排列,与整个区域长长的曲线形状相得益彰,同时构建了设计精巧的千回百转的内部展览空间,与艺术品保持了一种优雅的关系。卢浮宫朗斯分馆的外立面由经过阳极化处理的抛光铝建造,包裹着室内空间,外部轮廓隐隐约约、模糊不清。当人们漫步经过时,在不同的自然景观和可用光线条件下,会给人不同的印象感受。

主展览大楼位于入口大厅的侧面,大楼内一侧是大画廊,另一侧是临时展厅。通过入口大厅可以到达较低一层,这一层包含存储区和艺术品修复区。这样,博物馆也向公众开放了其"大后方"区域。

公园中,行政办公楼和餐馆是两座独立的建筑,把城市与博物馆连为一体。博物馆的入口位于前矿场的中心位置,也就是进入场地的入口,地势从保罗伯特街开始渐渐提升。建筑内的透明区域使游客可以一览周围绿树和朗斯市的优美风光。入口所在的位置既可以把整座博物馆的景色一览无余,其玻璃和铝质金属表面又可以把公园的全景揽入怀中。设计师把入口区域设计为一个上空体量,成为周围景观的一部分,人们在任何地方都可见到它。从主要的北入口到达博物馆的游客可以到此位置,从位于东面的绿地区域和西面树林到达博物馆的游客也可以到此。这处大型的透明区域非常宽敞,为到此的游客提供了多样化的功能区域。各个功能空间犹如玻璃"气泡",漂浮在大厅里,主要提供与公众相关的服务,为游客的个人活动提供空间。通往博物馆地下一层的入口位于迎接大厅的中央,邀请游客进入艺术品存储区和有洗手间和化妆间的服务区。

大厅的顶部洞口体现了大厅所展现的几何主题,位于平板上的右侧洞口将光线引入较低层。天花板由颜色非常浅的穿孔铝板覆盖,用来反射自然光线,如同漂浮在整个下面空间。立面是整面的落地双层保温玻璃,呈开间结构,装有卷动遮阳百叶窗。

Louvre-Lens Museum

In keeping with a desire to maintain the openness of the site and to reduce the ascendancy of this large project, the building was broken down into several spaces. Through their size and layout, which follow the gradual changes in terrain elevation, the buildings achieve a balance with the scale of the site and the shape of the paths, landscape features evoking its mining history.

In order to visually and physically open up the site, the main glassed area features a hollow in the core of the building. This delicate glass box serves as an entry hall to the museum and is a genuine public space for the city of Lens. It is transparent and opens up to several directions of the site, and it can be crossed

plant stratum		dynamic culture	
A	promenades lisières	I	bâtiment-clairière
A1	chemin creux	II	prairie estrade
A2	front des colonisateurs	III	grande esplanade
A3	haut du remblai	IV	pré
B	parc pionnier	V	plateforme est
C	jardins du louvre	VI	grande percée
a	seuil parvis	VII	bande active
b	terrasse des robiniers	VIII, IX, X	signal
c	terrasse du midi		
d	carré des arts vivants		
e	terrasse des arts		

mining base
1 puit 9
2 haut du remblai
3 terrasse Devocelle
4 grève des temils
5 front des colonisateurs
6 grand cavalier haut
7 chemin creux

植物地层 plant stratum   动态文化 dynamic culture   采矿基地 mining base

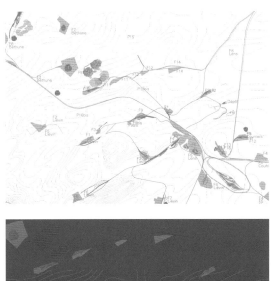

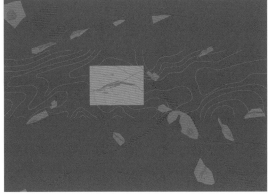

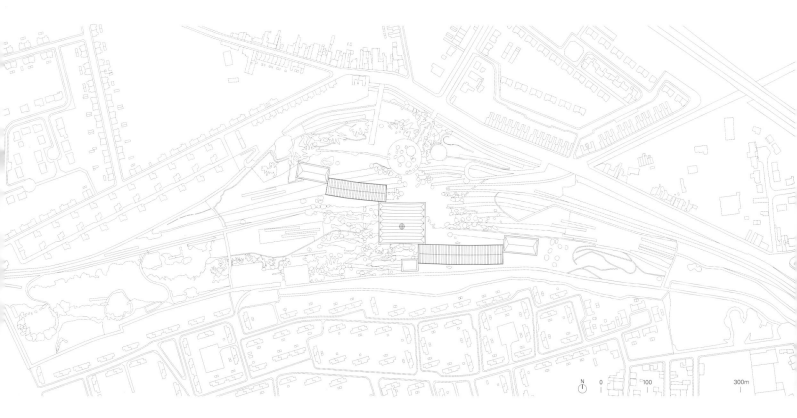

建筑师：Kazuyo Sejima, Ryue Nishizawa
项目团队：Yumiko Yamada, Yoshitaka Tanase, Louis-Antoine Grégo, Rikiya Yamamoto, Kohji Yoshida, Lucy Styles, Erika Hidaka, Nobuhiro Kitazawa, Bob Van den Brande, Arrate Arizaga Villalba, Guillaume Choplain, Osamu Kato, Naoto Noguchi, Shohei Yoshida, Takashige Yamashita, Takashi Suo, Ichio Matsuzawa, Andreas Krawczyk, Angela Pang, Jonas Elding, Sam Chermayeff, Jeanne-Francois Fischer, Sophie Shiraishi
博物馆技术：Studio Adrien Gardere, Imrey Culbert
景观设计：Mosbach Paysagistes
MEP和机构工程师：Betom Ingénierie　能源与舒适性理念设计：Transplan
环境设计工程师：Hubert Penicaud
结构概念顾问：Sasaki & Partners
结构与立面工程：Bollinger & Grohmann
艺术与自然光线设计：Arup
摄影师：©Iwan Baan (courtesy of the architect) - p.26~27, p.32 top, p.39
©Philippe Chancel (courtesy of the architect) - p.37, p.38
©Hisao Suzuki (courtesy of the architect) - p.30~31, p.34~35 (except as noted)

through to get to different quarters of the city.

The project avoids the strict, rectilinear shapes that would have conflicted with the subtle character of the site, as well as of free shapes that would have been overly restrictive from the perspective of the museum's internal operations. The slight inflection of the spaces is in tune with the long curved shape of the site and creates a subtle distortion of the inner areas while maintaining a graceful relationship with the artwork. The spaces are contained by a facade of anodized, polished aluminum that reverts a blurred and fuzzy image of the site's contours, and reflections change as one strolls by depending on the landscaping and available light.

The main exhibition buildings flank the entry hall, the Grand Gallery on one side and the temporary exhibition hall on the other. The entrance hall leads to a lower level that contains storage spaces and artwork restoration areas. The museum thus opens its rear areas to the public.

In the park, two free-standing buildings house the administration offices and the restaurant, linking the museum to the city. The entry to the museum is located in the center of the former pit and is the historical access to the site, rising gently from Paul Bert Street. The transparent areas in the building provide views of the surrounding wood and the city of Lens. This entry point provides a perspective of the entire building and of the panorama over the park reflected in the glass and aluminum surfaces. The entry area was designed as a void that is part of the landscape and visible from everywhere. It takes visitors to arrive at the museum from the main north entrance, as well as from the grassy areas to the east and the wood from the west. This large, transparent area is an ample space within which diverse functional areas exist for the museum's visitors. The glass "bubbles" seem to float within the interior of the hall. They are primary for public-related functions and provide areas for individual experiences. Access to the first lower level of the building is at the center of the hall, inciting visitors to enter the art storage area and the services area containing washrooms and dressing rooms.

Openings overhead reflect the geometric themes present in the hall, and the right openings in the slab direct light to the lower level. The ceiling is covered with sheets of perforated aluminum of a very light color, reflecting natural light and drifting over the entire underside. The facades are large, full-height glass bays that are double insulated. A system of roll-down shades provides protection from the sun.

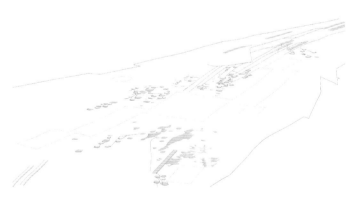

项目名称：Musée du Louvre-Lens
地点：Lens, France
总表面积：28,000m²　展区面积：7,000m²
服务区面积：6,000m²　大厅面积：36,000m²　公园面积：200,000m²
造价：EUR150,000,000　设计时间：2005—2009　施工时间：2009—2012.12

北立面 north elevation

1. sunshade grille
2. double glazing glass sky-light 10+(8+8)mm
3. interior movable louver
4. paint on steel T-beam w250t=20xh600-1,100t=10mm
5. aluminum honeycomb panel t=2+20+1mm
6. insulation t=140mm
7. concrete t=280mm
8. aluminum honeycomb panel t=1+(1+18+1)mm
9. polished concrete screed t=150mm
10. insulation 90mm (floor heating)
11. structural concrete t=240mm

1. kalzip
2. acoustic board
3. perforated metal t=1.8mm
4. fascia
5. sunshade rolled screen
6. double glazing glass 10+(8+8)mm
7. mullion
8. column
9. galvanized grating
10. polished concrete screed t=130mm
11. insulation 90mm (floor heating)
12. 3-I structural concrete t=250mm

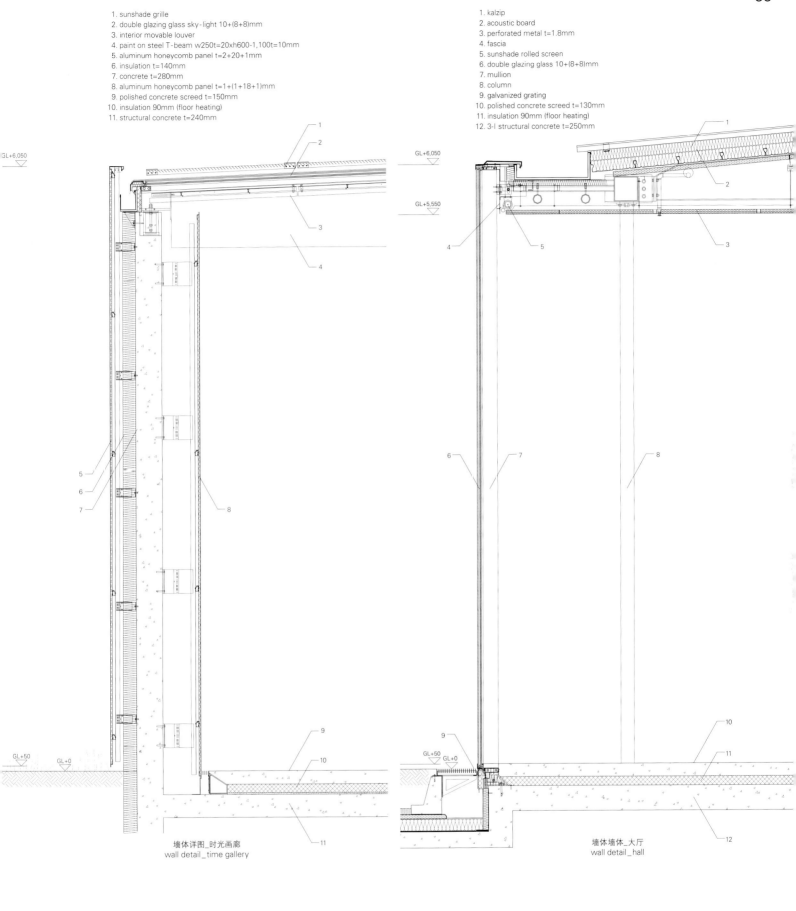

墙体详图_时光画廊
wall detail_time gallery

墙体墙体_大厅
wall detail_hall

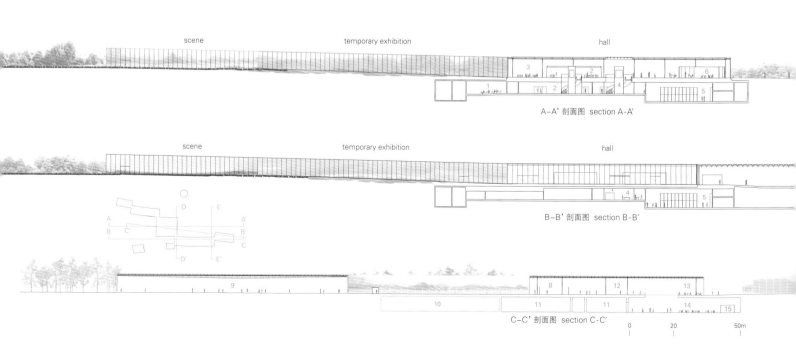

A-A' 剖面图 section A-A'

B-B' 剖面图 section B-B'

C-C' 剖面图 section C-C'

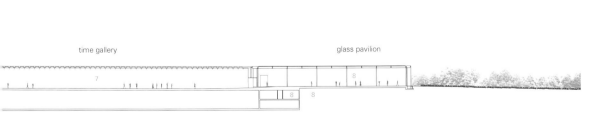

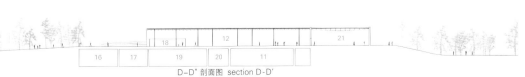

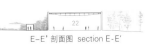

1. petit auditorium
2. centre de ressources
3. cafétéria
4. espace découverte
5. reserves d'oeuvres d'art
6. espace adhérents
7. bulles mécène
8. atelier
9. présentations renouvelées
10. réserves louvre
11. réserves pérenne visibles et visitables
12. accueil général du public
13. salon de cafeteria
14. mediathéque
15. rangement mobilier
16. stockages divers
17. aire de chargement et dechargement materiels/produits
18. librarie boutique
19. rangement chariots et matériel de manutention local de recharge des batteries de chariots
20. stockage des caisses
21. le cheminement dans le temps espace de découverte et
22. d'approfondissement

time gallery

glass pavilion

D-D' 剖面图 section D-D'

E-E' 剖面图 section E-E'

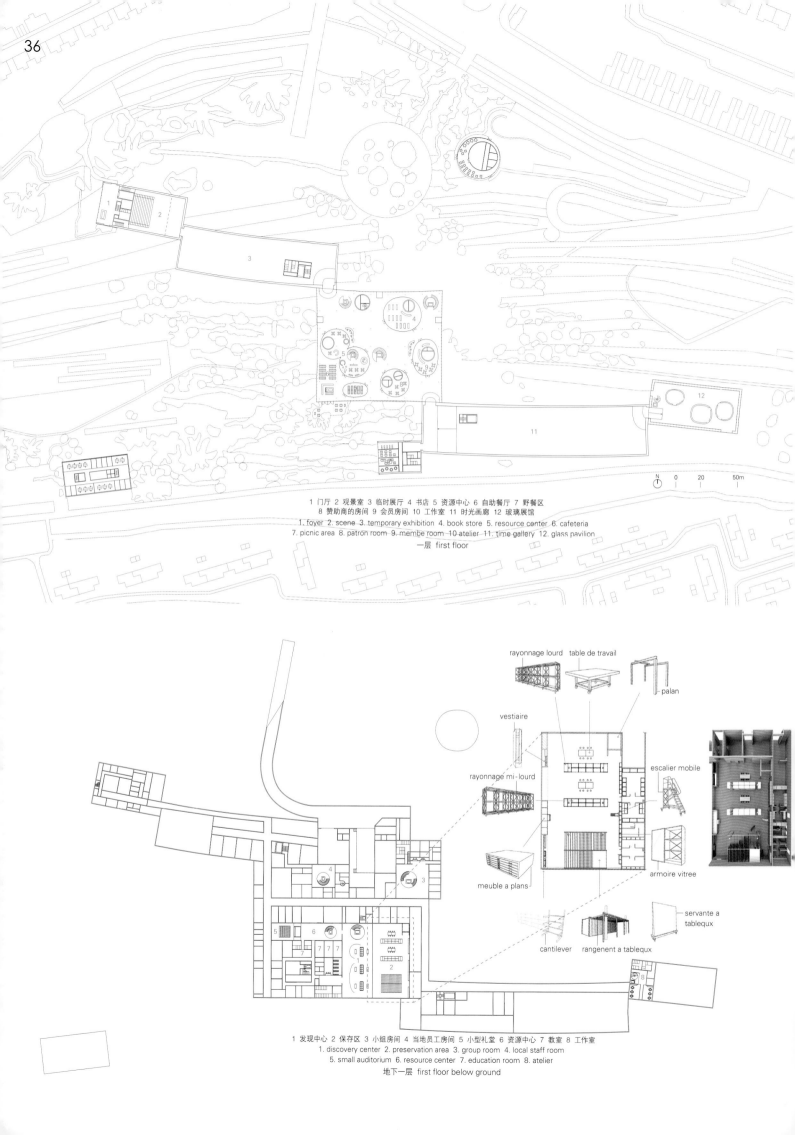

1 门厅 2 观景室 3 临时展厅 4 书店 5 资源中心 6 自助餐厅 7 野餐区
8 赞助商的房间 9 会员房间 10 工作室 11 时光画廊 12 玻璃展馆
1. foyer 2. scene 3. temporary exhibition 4. book store 5. resource center 6. cafeteria
7. picnic area 8. patron room 9. membe room 10. atelier 11. time gallery 12. glass pavilion
一层 first floor

1 发现中心 2 保存区 3 小组房间 4 当地员工房间 5 小型礼堂 6 资源中心 7 教室 8 工作室
1. discovery center 2. preservation area 3. group room 4. local staff room
5. small auditorium 6. resource center 7. education room 8. atelier
地下一层 first floor below ground

LA GALERIE DU TEMPS

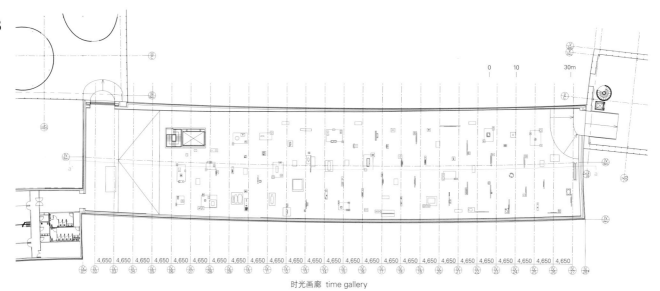

时光画廊 time gallery

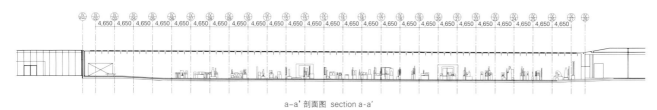

a-a' 剖面图 section a-a'

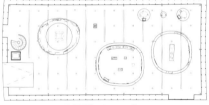

玻璃展馆 glass pavilion

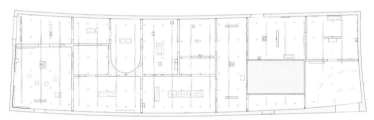

临时展厅 temporary exhibition

佩洛特自然和科学博物馆
Morphosis Architects

博物馆,是集体社会经验和文化表达方式的电枢,展现给人们的是以全新的方式来诠释世界。博物馆充满知识,保存信息和传递思想;激发好奇心,提高意识,创造交流机会。作为教育工具和体现社会变化的工具,博物馆能够帮助人们对所生活的世界进行理解。

随着全球环境面临着越来越严峻的挑战,更广泛地了解自然系统之间的相互依赖对人类的生存和进化来说越来越有必要,而自然和科学博物馆在加深人们对这些复杂的自然系统的理解方面至关重要。

位于胜利公园的佩洛特自然和科学博物馆为博物馆创造了一个与众不同的身份,增强了其在达拉斯市的重要性,丰富了城市不断发展的文化肌理。博物馆旨在吸引更广泛意义上的参观者,激励年轻人的思想,激发参观者对日常生活的好奇心和求知欲,带给参观者一份难忘的体验。这份体验将铭记在参观者的心中,并最终加深参观者和社会对自然和科学的理解。

这座世界级的博物馆提供了具有互动性的参观环境,参观者可以沉浸其中,以充分激发了人们对科学的认识。博物馆设计摒弃了博物馆只是作为展品的中立性背景的这一理念,使博物馆本身成为进行科学教育的一个有效工具,通过集建筑、自然和科技于一身,佩洛特自然和科学博物馆既展示了科学原理,又激发了人们对周围自然环境的好奇心。

在这个城市中,身临其境的大自然体验始于参观者接近博物馆。靠近博物馆,映入参观者眼帘的是两处德克萨斯州原生生态系统的代表:德州本地长有大树冠的树木组成的树林和体现本地沙漠条件下旱生植物群落的旱生园艺景观阶地。体现旱生园艺景观的阶地坡度缓缓抬高,与博物馆的标志性石屋顶连为一体。整座建筑的主体被设计成一个大立方体,浮在经过景观美化的基座上。面积为一英亩(4046m²)的如波浪起伏般的屋顶景观由岩石和本地耐旱草本植物构成,反映了达拉斯市的本土地质情况,向参观者展示了一个随着时间推移而自然演变的生命系统。

这两处生态系统交汇处就是主入口广场,是参观者参加聚会和举办活动的区域,也是达拉斯市的一个户外公共空间。从广场处,参观者可以沿着缓缓提升的景观屋顶穿过一个相对狭小空间进入到宽敞的入口大厅。大厅的天花板造型也是波状起伏,充分表现了外部景观表面的动态与活力,模糊了内部和外部之间的界线,将自然与人造制品融为一体。

经过入口相对狭小的空间时,参观者的目光会被上面吸引,看到高耸而宽敞、充满自然光线的中庭,这是博物馆主要的交通空间,光线充足,楼梯、自动扶梯和电梯都在此处。

在地面一层,参观者可以乘坐一系列自动扶梯通过中庭到达博物馆的最高一层。参观者可以到高高的位于城市上空的全部用玻璃围挡的阳台上,鸟瞰达拉斯市中心。从这个空中阳台开始,参观者可以穿过画廊沿着螺旋通道顺时针缓缓而下。这处动态的空间行进过程会给参观者带来震撼内心的体验,使参观者沉浸于博物馆的建筑和自然环境之中。从顶层缓缓而下的通道穿过博物馆不同的画廊,一会儿与博物馆的主要交通中枢——中庭交织在一起,一会儿又独立开来,使参观者既可以融

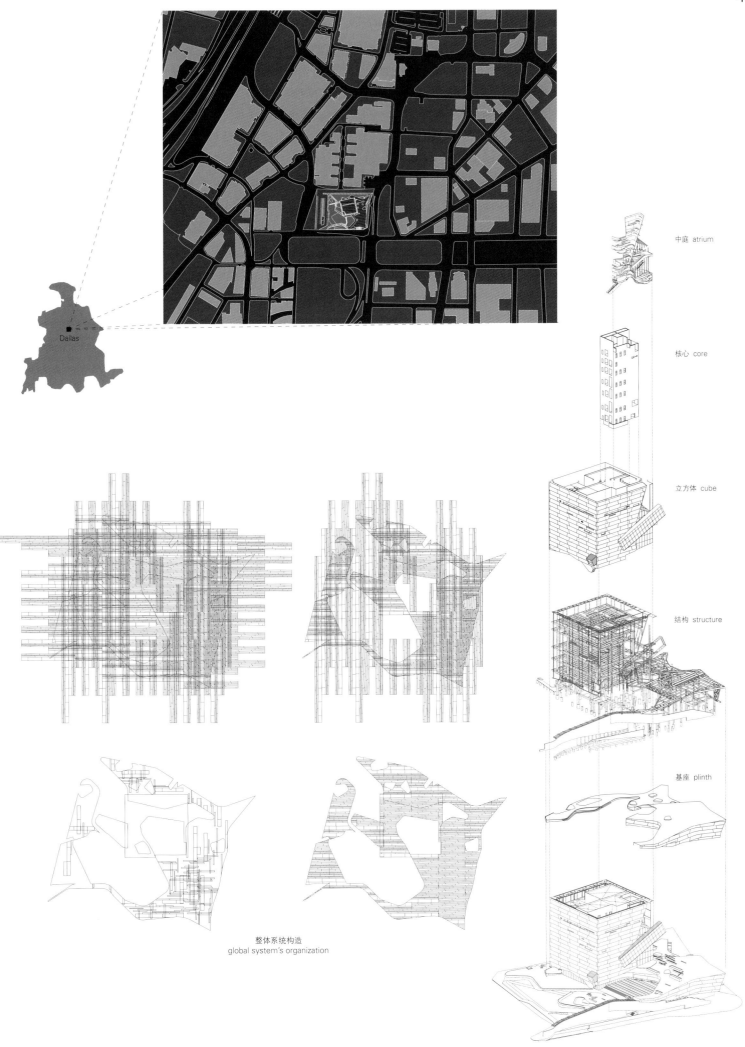

入博物馆的内部世界,又可以保持着与博物馆外部城市生活的联系。参观者也成为博物馆建筑的一部分,因为博物馆东面一角面向达拉斯市的市中心,人们对参观者在博物馆内部的活动可以一览无余。因此可以说,博物馆从根本上来说就是一座公共建筑,它对这座城市开放,属于这座城市并激活这座城市;最终,公众成为博物馆不可缺少的组成部分,如同博物馆成为这座城市不可缺少的组成部分。

Perot Museum of Nature and Science

Museums, armatures for collective societal experience and cultural expression, present new ways of interpreting the world. They contain knowledge, preserve information and transmit ideas; they stimulate curiosity, raise awareness and create opportunities for exchange. As instruments of education and social change, museums have the potential to shape understanding of the world in which people live.

As global environment faces ever more critical challenges, a broader understanding of the interdependence of natural systems is becoming more essential to survival and evolution. Museums dedicated to nature and science play a key role in expanding understanding of these complex systems.

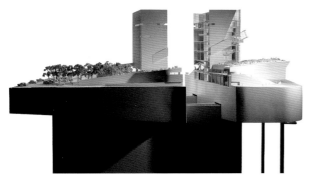

1. F.S.T., b1/a-005.6 w/ concrete formed end panel
2. concrete 4"x14ga steel studs @16"oc-attach to novum steel outriggers
3. concrete 2 1/2"x14ga back to back steel studs-attach to novum steel outriggers
4. concrete sealant and backer rod, both sides, TYP
5. painted steel attachment plate
6. 14ga PTD. metal panel color to match duranar "pewter gray"
7. F.S.T., c4/a-005.7
8. gypsum board on metal stud as required
9. PTD "F" shape mtl reveal molding w/mudable face
10. CMP-see finish schedule upturn finished panel @exposed edges, typ stop perforation pattern @bottom edge of tile
11. concrete 14ga Z clips @24" oc-typical
12. tube steel glazing channel support framing as REQ'D by novum
13. 3" metal deck
14. structural steel w/intumescent fire proofing
15. PTD metal reveal trim w/mudable face
16. steel structure w/fireproofing
17. steel outrigger for glass channel by novum
18. MTL flashing, PTD to match mtl panel w/drip edge and tucked into glazing pocket
19. shim as REQ'D
20. concrete formed end panel
21. concrete WP
22. F.S.T., b1/a-005.6 @soffit
23. abuse-resistant GYP. BD w/non-slip textured surface
24. 3/4" plywood (fire treated)
25. stud spacing to be engineered for walkable surface

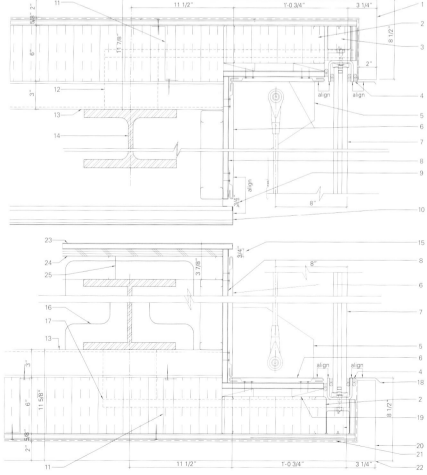

管状外挂电梯详图 escalator tube detail

项目名称：Perot Museum of Nature & Science
地点：2201 N. Field Street, Dallas, Texas, USA
建筑师：Morphosis Architects
设计总监：Thom Mayne
项目委托人：Kim Groves
项目经理：Brandon Welling
项目建筑师：Arne Emerson
项目设计师：Aleksander Tamm-Seitz
项目团队：Natalia Traverso Caruana, Paul Choi, Kerenza Harris, Sal Hidalgo, Andrea Manning, Aaron Ragan, Scott Severson, Martin Summers, Jennifer Workman
合作建筑师：Good Fulton & Farrell
结构工程师：Datum Engineers
景观建筑&场地可持续性：Talley Associates
用地面积：19,000m² 有效楼层面积：16,722m²
竣工时间：2013
摄影师：©Roland Halbe

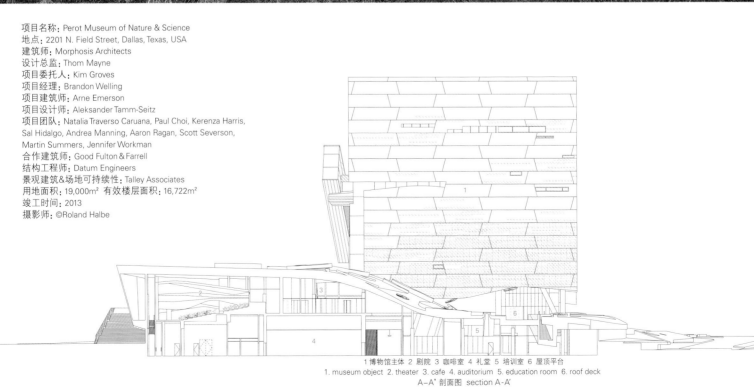

1 博物馆主体 2 剧院 3 咖啡室 4 礼堂 5 培训室 6 屋顶平台
1. museum object 2. theater 3. cafe 4. auditorium 5. education room 6. roof deck
A-A' 剖面图 section A-A'

Perot Museum of Nature and Science in Victory Park creates a distinct identity for the museum, enhances the institution's prominence in Dallas and enriches the city's evolving cultural fabric. Designed to engage a broad audience, invigorate young minds, and inspire wonder and curiosity in the daily lives of its visitors, the museum cultivates a memorable experience that persists in the minds of its visitors and that ultimately broadens individuals' and society's understanding of nature and science.

This world class facility inspires awareness of science through an immersive and interactive environment that actively engages visitors. Rejecting the notion of museum architecture as neutral background for exhibits, the new building itself is an active tool for science education. By integrating architecture, nature, and technology, the building demonstrates scientific principles and stimulates curiosity in our natural surroundings.

The immersive experience of nature within the city begins with the visitor's approach to the museum, which leads through two native Texas ecologies: a forest of large native canopy trees and a terrace of native desert xeriscaping. The xeriscaped terrace gently slopes up to connect with the museum's iconic stone roof. The overall building mass is conceived as a large cube floating over the site's landscaped plinth. An acre of undulating roofscape comprised of rock and native drought-resistant grasses reflects Dallas's indigenous geology and demonstrates a living system that will evolve naturally over time.

The intersection of these two ecologies defines the main entry plaza, a gathering and event area for visitors and an outdoor public space for the city of Dallas. From the plaza, the landscaped roof lifts up to draw visitors through a compressed space into the more expansive entry lobby. The topography of the lobby's undulating ceiling reflects the dynamism of the exterior landscape surface, blurring the distinction between inside and outside, and connecting the natural with the manmade.

Moving from the compressed space of the entry, the visitor's gaze is drawn upward through the soaring open volume of the sky-lit atrium, the building's primary light-filled circulation space, which

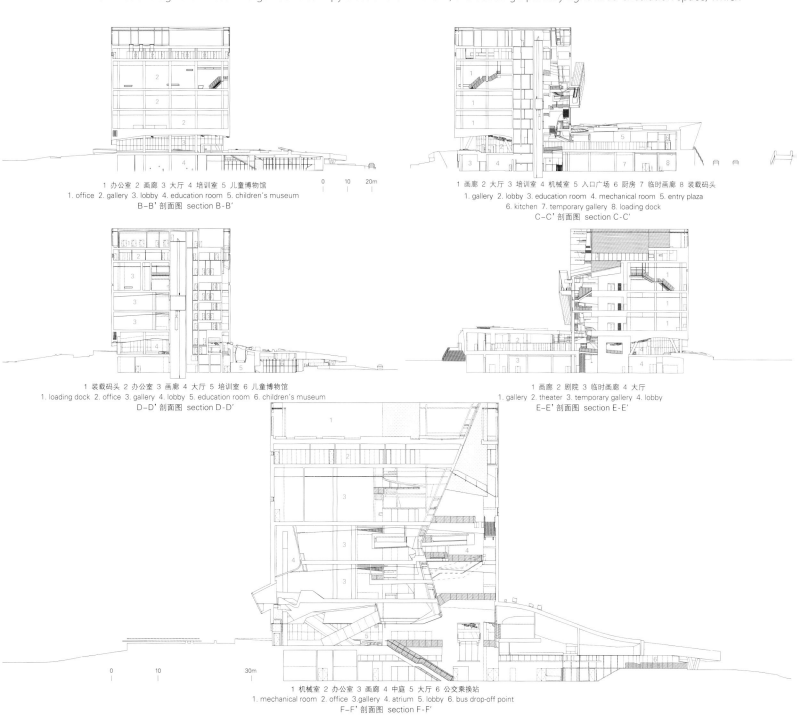

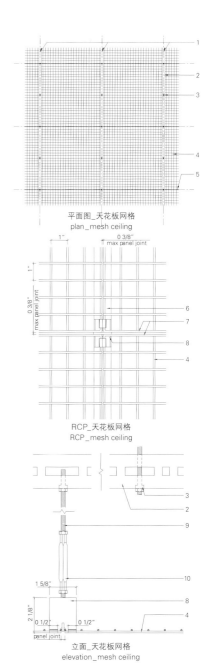

平面图_天花板网格
plan_mesh ceiling

RCP_天花板网格
RCP_mesh ceiling

立面_天花板网格
elevation_mesh ceiling

through lobby ceiling

1. bracing above ceiling as required with cables: coordinate with mep and structure above
2. painted metal conduit feed tight to exterior threaded rod
3. typical perforations diam 1/4", 1/2", 3/4", 1 1/2", 2 1/2" REF digital model
4. threaded rod suspended and braced as required with cables. powder coat black
5. turnbuckle powder coat black
6. rigidly attach threaded rod to tube steel frame
7. joint line between shells
8. min 1/8" fiberglass with smooth finish and custom color on exterior
   - interior surfaces to be painted opaque white
   - access panel to be frameless with concealed hinges at approved location
9. 2" max continuous gap at the edge of ceiling panel TYP all sides of element
10. steel light fixture support clip by element mfgr for suspension of fixture within element, coordinate with fixture MFGR requirements and electrical
11. light fixture
12. 1x1 structural tube steel frame: fabrication and engineering by element manufacturer.
    - all internal structure to be matte white
13. ground support locations

mesh ceiling suspension system

1. unidirection suspension system. align with mesh panel joints below
2. cold-rolled steel "U" channel support grid. provide powder coat finish
3. suspension from structure above as required 1"x1" cell welded wire mesh panel.
4. provide powder coat finish
5. panel joint TYP
6. cold-rolled steel "U" channel support grid above.
7. continuous edge band TYP all mesh panels
8. custom steel ceiling suspension clip above. provide powder coat finish
9. all-thread rod support as required. provide powder coat finish
10. turnbuckle. provide powder coat finish

立面_可分开的隔舱
elevation_POD

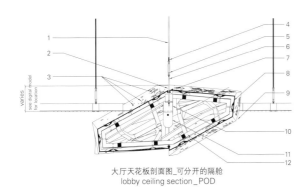

大厅天花板剖面图_可分开的隔舱
lobby ceiling section_POD

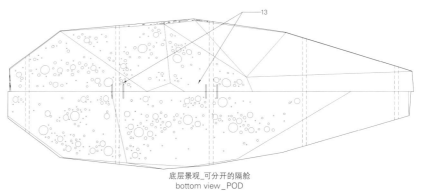

底层景观_可分开的隔舱
bottom view_POD

1. wall type D2
2. interior finish per schedule
3. precast concrete facade, facade system type 4 (rated condition)
4. 2hr fire rated floor assembly
5. concrete precast substructure
6. sprayed on waterproofing over EXT. sheathing over CFMF
7. lighting fixture re: a-151.0a
8. wire mesh ceiling at tower lobby and exterior
9. exposed concrete column
10. tension supported glazing
11. tension rod
12. trench heater per mechanical
13. tension rod embed
14. roof type no 3 or 4, ref. a-006.0
15. fireproofed steel, TYP.
16. precast support steel
17. precast concrete facade panel
18. gypsum wall over CFMF
19. theater stairs beyond
20. steel precast support; REF. structural
21. concrete. black sprayed insulation coating
22. exterior soffit

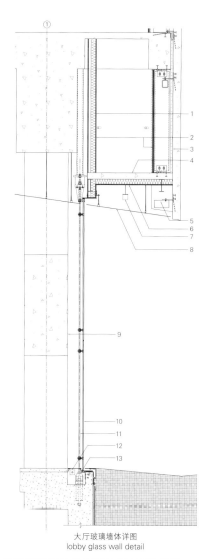
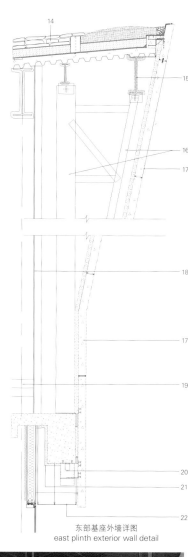

大厅玻璃墙体详图
lobby glass wall detail

东部基座外墙详图
east plinth exterior wall detail

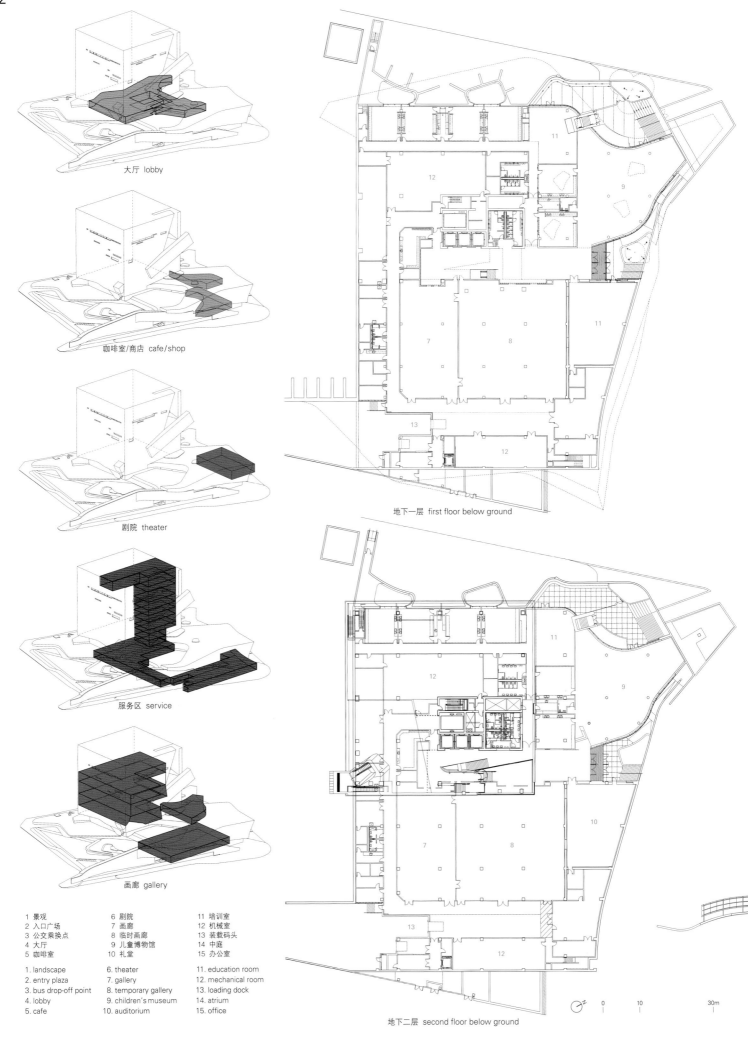

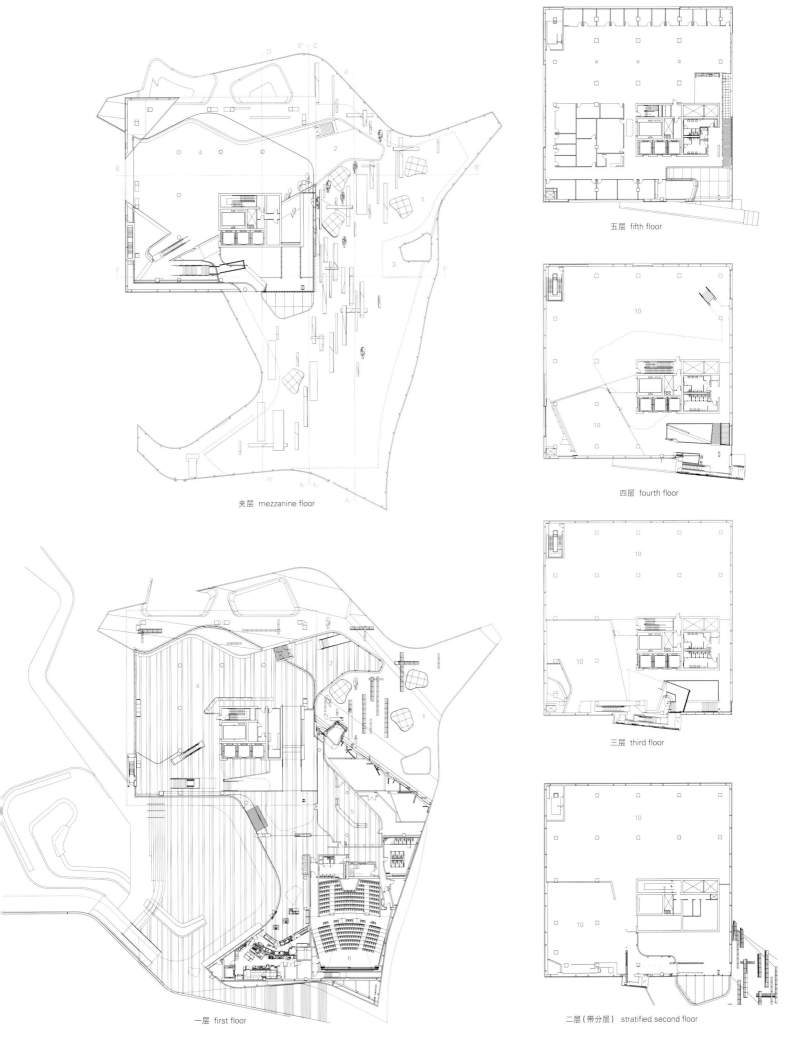

夹层 mezzanine floor

一层 first floor

五层 fifth floor

四层 fourth floor

三层 third floor

二层（带分层） stratified second floor

houses the building's stairs, escalators and elevators.

From the ground floor, a series of escalators bring patrons though the atrium to the uppermost level of the museum. Patrons arrive at a fully glazed balcony high above the city, with a bird's eye view of downtown Dallas. From this sky balcony, visitors proceed downward in a clockwise spiral path through the galleries. This dynamic spatial procession creates a visceral experience that engages visitors and establishes an immediate connection to the immersive architectural and natural environment of the museum. The path descending from the top floor through the museum's galleries weaves in and out of the building's main circulation atrium, alternately connecting the visitor with the internal world of the museum and with the external life of the city beyond. The visitor becomes a part of the architecture, as the eastern-facing corner of the building opens up towards downtown Dallas to reveal the activity within. The museum, is thus, a fundamentally public building – a building that opens up, belongs to and activates the city; ultimately, the public is as integral to the museum as the museum is to the city.

大玛雅文明博物馆
Grupo Arquidecture

大玛雅文明博物馆以当代建筑语言阐释了玛雅人所崇拜的东西而非玛雅人所建造的东西。在对玛雅文明的追寻中,人们会经常发现这样一个象征符号,一个玛雅文化宇宙观中的关键元素:木棉树,玛雅人的圣树,其根部会直达地下世界并适应那里的环境,树叶覆盖的树枝伸向天空,到达人类超然存在的地方,而繁茂枝叶形成的树荫下——树干,是生命的所在,是人们日常活动的地方。

在这一理念下,世界是由三块石头和木棉树建造的,建筑师在设计时把功能需求和所需的不同的功能空间通盘考虑,其结构设计理念不仅满足建筑的物理承载要求,也满足了为博物馆所有区域提供保障和支持的其他基础设施工程要求。

博物馆的展品收藏区、转运区、重大考古发现和研究区以及拥有260个停车位的停车场都位于"木棉树的根部"层。

拾级而上,到达"木棉树的树干"位置,这里有博物馆主大厅、售票处、私人物品保存区、2000m^{2}的永久性展区和600m^{2}的巡回展区、公共关系办公室、儿童看护中心、带露台的餐厅、纪念品商店以及一个露台酒吧。

行政办公区域坐落在"木棉树的枝叶"里,这里还有高清宽幅影院。电影院包括多功能大厅以及可以举行多种多样艺术和文化活动的演出设施。

建筑师认为可持续发展应该成为任何项目不可分割的一部分。这里,本项目从如下方面阐释可持续发展:

环境

博物馆位于梅里达市北部,即城市重要分中心的核心位置,设计致力于寻求自然通风和光线,使用被动系统实现能源效益和环境舒适。博物馆主大厅是一个中空的核心区,被树影婆娑的"木棉树"覆盖,将建筑的不同楼层连为一体。

建筑师设计了一座全面的博物馆,努力做到使每一处空间适合于每一个用户;为了让所有参观者参观博物馆时享有同样的尊贵,台阶处两侧设计了相互交织的坡道,还有和人行道位于同一水平位置的电梯,以及位于停车场内的电梯,盲文信号指示、走廊和卫生间内所有设施都便于老年人或残疾人使用,参观途中设有休息区,还有为提高员工生活质量而设置的空间,机器和设备间的设计兼顾机器和操作机器的人两个因素,把员工与参观者置于同样重要的地位。

经济

设计中遵循重要的经济合理性指导方针,具体体现在功能区域直接使用被动系统,选择当地的建筑材料和建筑系统,只雇佣当地和本区域的建筑公司和员工。同时,做到基础设施工程项目设计最优化,从操作和维护两方面达到资源的最佳利用。

Great Museum of the Maya Civilization

Great Museum of the Maya Civilization is a building with a contemporary expression about what the Mayans worshiped rather than the Mayans built, in this search we found a recurrent symbol, a key element in the cosmic vision of Mayan Culture: Ceiba, the sacred tree, whose roots penetrates and conforms the underworld, the trunk's level lays down where life and daily activities take place underneath the shade of its frond which spreads its branches up to the sky and human transcendence.

项目名称：Great Museum of the Maya Civilization
地点：Calle 60 Norte, No. 299-E, Colonia Revolución Ex Cordemex, Mérida, Yucatán, Mexico
建筑师：Ricardo Combaluzier, Enrique Duarte, William Ramírez, Josefina Rivas
设计团队：Luis De La Rosa, Alma Villicaña, Carlos Guardián, Mariana Farfán, Julio Rosas, Aída Ordóñez, Fabián Rosas, Ricardo Combaluzier
结构设计：Rodolfo Pascacio
电气工程师：Rafael Sánchez
景观建筑师：Tania Domínguez
甲方：Patronato de las Unidades de Servicios Culturales y Turisticos del Estado de Yucatan, Gorbierno del Estado de Yucatan
用地面积：23,254m² 总建筑面积：22,600m²
有效楼层面积：11,800m²
竣工时间：2010
摄影师：©David Cervera (courtesy of the architect) (except as noted)

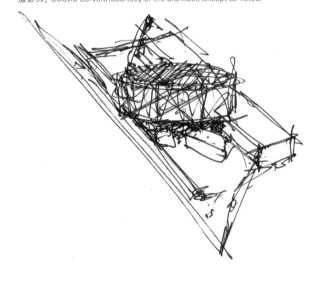

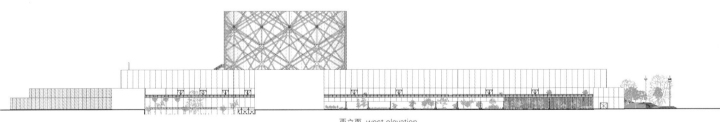

东立面 east elevation

西立面 west elevation

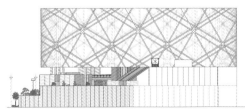

北立面 north elevation

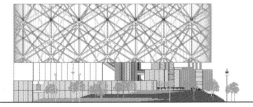

南立面 south elevation

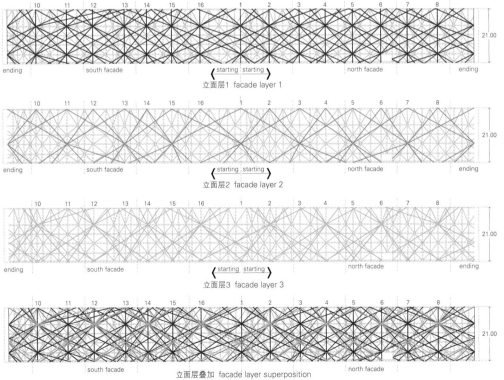

立面层1 facade layer 1

立面层2 facade layer 2

立面层3 facade layer 3

立面层叠加 facade layer superposition

1. museum collections
2. collection transit area
3. staff area
4. multimedia room
5. parking
6. mechanical equipment room
7. permanent exhibition room
8. travelling exhibition room
9. main lobby
10. ticket offices
11. souvenir shop
12. restaurant
13. public relations office
14. childcare center
15. restaurant with terrace
16. family restrooms
17. bathroom facilities
18. terrace bar
19. high-definition large format cinema
20. cafe
21. multi-purpose hall
22. administrative offices

1 博物馆收藏区
2 收藏过渡区
3 员工区
4 多媒体室
5 停车场
6 机械设备室
7 永久性展览室
8 巡回展览室
9 主大厅
10 售票室
11 纪念品商店
12 餐厅
13 公共关系办公室
14 儿童看护中心
15 带露台的餐厅
16 家庭休息室
17 浴室设施
18 露台酒吧
19 高清宽幅影院
20 咖啡室
21 多功能大厅
22 行政办公室

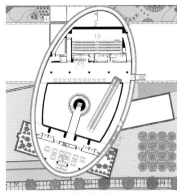

三层 third floor

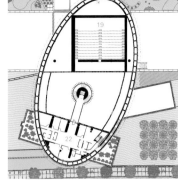

四层 fourth floor

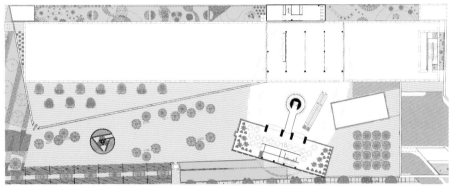

二层（带分层） stratified second floor

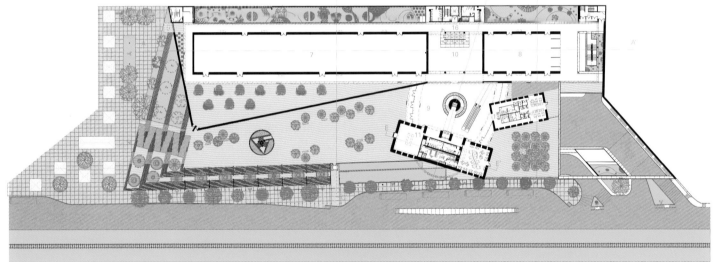

二层 second floor

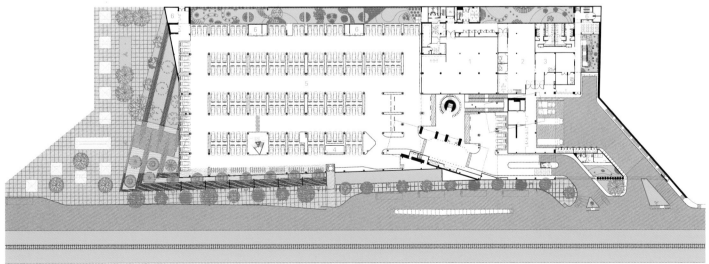

一层 first floor

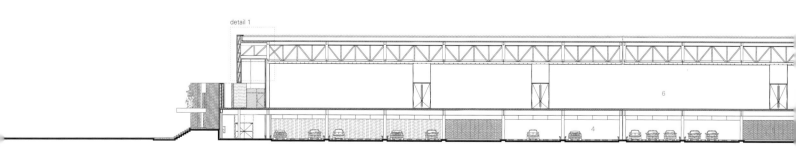

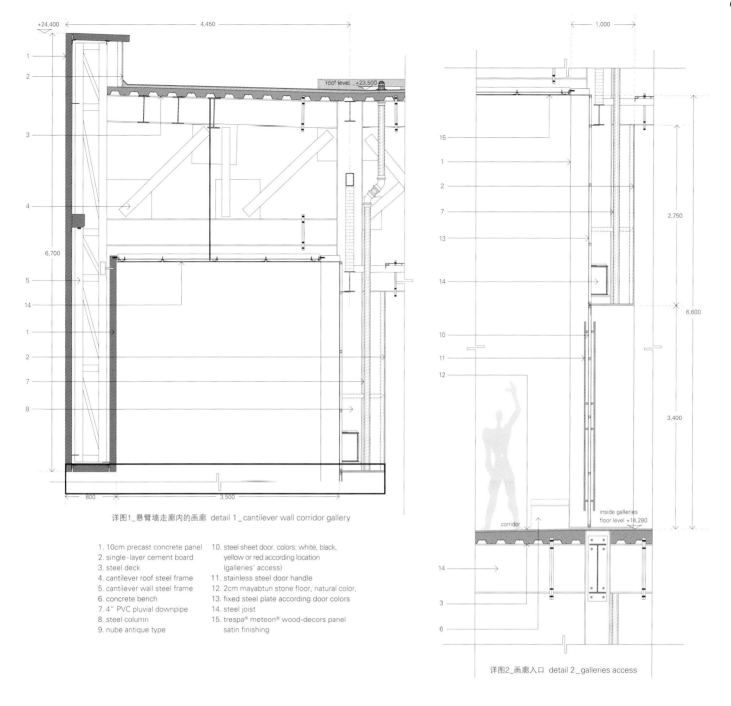

详图1_悬臂墙走廊内的画廊 detail 1_cantilever wall corridor gallery

1. 10cm precast concrete panel
2. single-layer cement board
3. steel deck
4. cantilever roof steel frame
5. cantilever wall steel frame
6. concrete bench
7. 4" PVC pluvial downpipe
8. steel column
9. nube antique type
10. steel sheet door. colors: white, black, yellow or red according location (galleries' access)
11. stainless steel door handle
12. 2cm mayabtun stone floor, natural color,
13. fixed steel plate according door colors
14. steel joist
15. trespa® meteon® wood-decors panel satin finishing

详图2_画廊入口 detail 2_galleries access

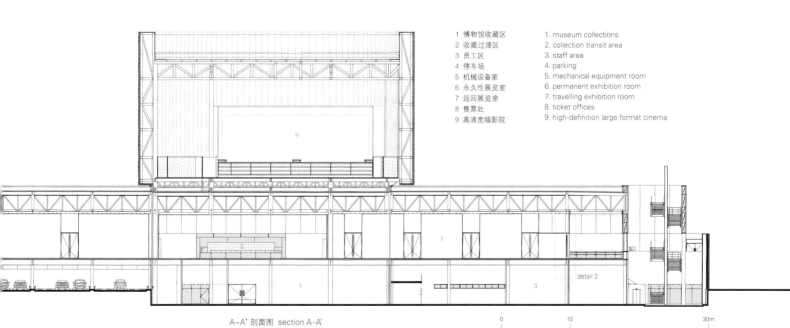

1 博物馆收藏区	1. museum collections
2 收藏过渡区	2. collection transit area
3 员工区	3. staff area
4 停车场	4. parking
5 机械设备室	5. mechanical equipment room
6 永久性展览室	6. permanent exhibition room
7 巡回展览室	7. travelling exhibition room
8 售票处	8. ticket offices
9 高清宽幅影院	9. high-definition large format cinema

A-A' 剖面图 section A-A'

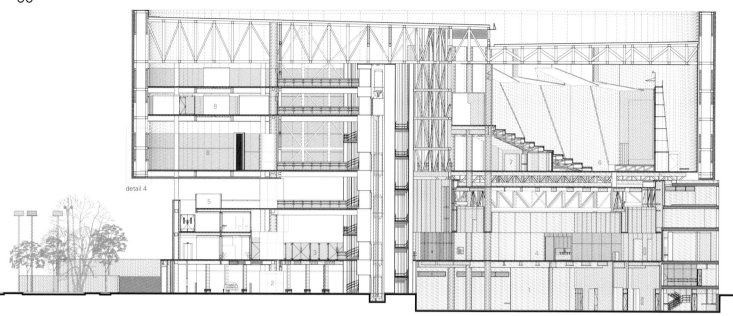

1 博物馆收藏区 2 停车场 3 主大厅 4 售票处 5 露台酒吧 6 高清宽幅影院 7 咖啡室 8 多功能大厅
1. museum collections 2. parking 3. main lobby 4. ticket offices 5. terrace bar 6. high-definition large format cinema 7. cafe 8. multi-purpose hall
B-B' 剖面图 section B-B'

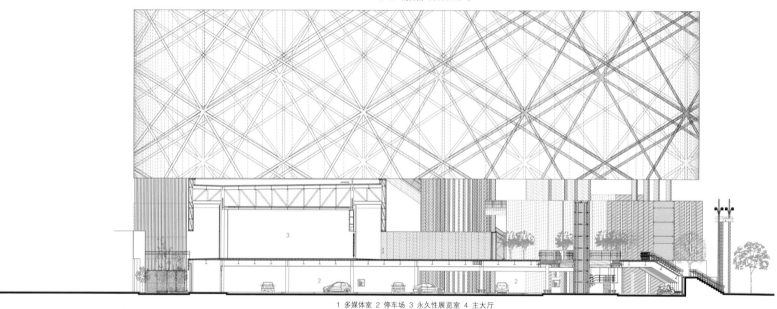

1 多媒体室 2 停车场 3 永久性展览室 4 主大厅
1. multi-media room 2. parking 3. permanent exhibition rooms 4. main lobby
C-C' 剖面图 section C-C'

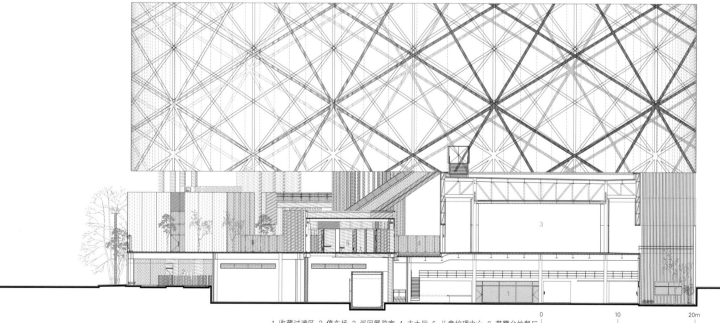

1 收藏过渡区 2 停车场 3 巡回展览室 4 主大厅 5 儿童护理中心 6 带露台的餐厅
1. transit collection area 2. parking 3. travelling exhibition room 4. main lobby 5. childcare center 6. restaurant with terrace
D-D' 剖面图 section D-D'

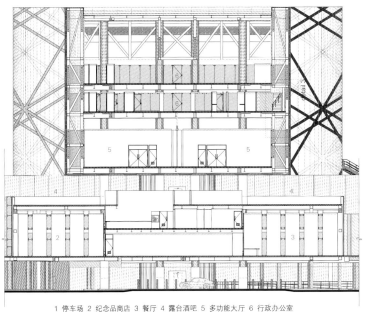

1 停车场 2 纪念品商店 3 餐厅 4 露台酒吧 5 多功能大厅 6 行政办公室
1. parking 2. souvenir shop 3. restaurant 4. terrace bar 5. multi-purpose hall 6. administrative offices
E-E' 剖面图 section E-E'

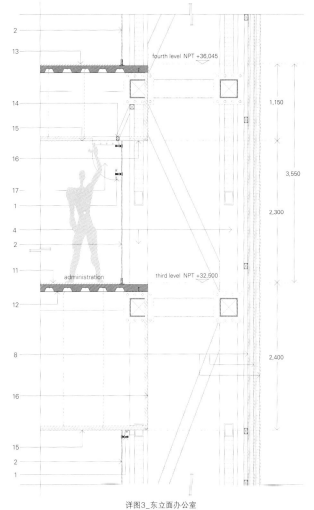

详图3_东立面办公室
detail 3_east facade offices

1. stainless steel heavy duty double arm spider fittings
2. 10mm tempered glass
3. 1 1/2" stainless steel handrail
4. elliptical steel frame structure
5. irving smooth steel grating step IS-02 type
6. red aluminum composite panel mounted on 2"×2" CFSF
7. door closer
8. trespa® meteon® HPL panel mounted on 4"×3" RHS
9. emergency exit double door
10. anti-panic door handle
11. tropical hardwood engineered
12. wood flooring
13. steel deck
14. concrete floor epoxy finishing
15. 3"×3" RHS frame to hold the glass fins gypsum base board ceiling and matte white latex paint
16. "L" shape cement board
17. ceiling and matte white latex paint 19mm tempered glass fins 50cm width

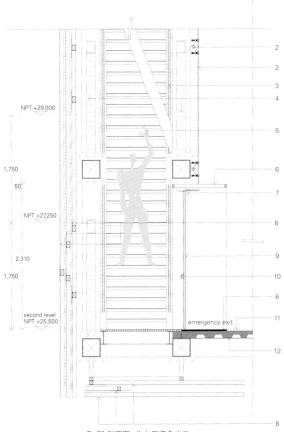

F-F' 剖面图_北大厅紧急出口
section F-F'_north hall emergency exit

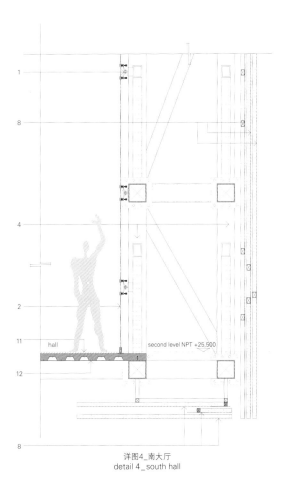

详图4_南大厅
detail 4_south hall

With this concept of the world's creation up from three stones and the Ceiba tree, the architects present the architectural design integrating the program needs and required spaces for different functional activities, the structural design concept that gives physical bearing to the building and to the other infrastructure engineerings shared out for nurturing and supporting all the museum's areas.

Museum collections, transit cellars, research and study areas of the great archaeological acquisition and a 260 parking space area are located at the "Ceiba's roots" level.

At the "Ceiba's trunk" level, up the perron, there contains the main lobby, ticket offices, personal belongings kept area, 2,000m² for permanent exhibition rooms and 600m² for travelling exhibitions, public relations office, childcare center, restaurant with terrace, souvenir shop and a terrace bar.

Executive and administrative offices are located inside the "Ceiba's frond", so is the high-definition large format cinema which includes performing arts facilities for various artistic and cultural activities as well as the multi-purpose hall.

The architects understand sustainability as an integral part of any project and in this meaning the aspects considered are:

Environment

Located at the heart of an important urban sub-center at north of Merida, the design raises seeking for natural air and light using passive systems to achieve energy benefits and environmental comfort. The main lobby covered and shaded by "the Ceiba" which holds a hollow core, joints the different floors of the building.

They designed a comprehensive museum, pretending to make suitable every space to every user: a twined ramp at the perron, a sidewalk-level elevator and the other in the parking lot, to yield universal access with equal dignity, braille signaling, all facilities in corridors and restrooms for the elderly or handicapped people, rest areas while taking the tour, spaces for workers to improve their life quality, machinery and equipment rooms designed for both, machines and people who operate them, bestowing to workers the same importance as visitors.

Economy

Significant economic rationality guidelines were followed, displayed in the design of functional spaces with direct use of passive systems and the selection of materials and building systems that privileged participation of local and regional companies and their employees, and the optimization in infrastructure engineerings projects achieves the best use of resources for the operation and maintenance. Grupo Arquidecture

宽容与记忆博物馆
Arditti + RDT Arquitectos

项目名称：Museum of Memory and Tolerance
地点：Mexico City, Mexico
建筑师：Arditti + RDT Arquitectos
建造商：Ideurban
结构工程师：Aguilar Ingenieros 设备工程师：TEDD
电气工程师：INSTATECH 机械工程师：Cyvsa
甲方：Fundación Memoria y Tolerancia AC
用地面积：1,100m² 面积：7,500m²
竣工时间：2010
摄影师：Courtesy of the architect

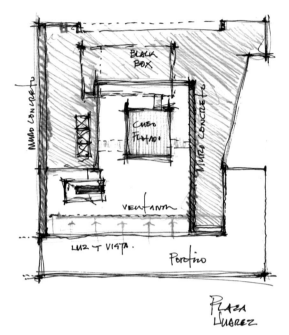

宽容与记忆博物馆将两部分,即对种族歧视激起的种族灭绝大屠杀的回忆(记忆部分)和这份记忆带给我们的不可饶恕,因而引导我们学会尊重他人、在多样化的世界里和平共处的遗产(宽容部分)进行结合。这座博物馆为墨西哥提供了在民主和多元文化的框架内为后代的发展进行研究的空间。

博物馆由钢筋混凝土和钢材混合建造,共7层(三个永久性展览层+四个临时或巡回展览层),构建于Legorreta+Legorreta建筑师事务所设计的华雷斯广场综合设施的一系列柱廊之上,这里原是"阿拉米达酒店"的旧址(已于1985年的地震中倒塌)。

Arditti+RDT建筑事务所深信,人道主义唯一的希望在于对后代的教育。博物馆的设计正是基于这一根深蒂固的理念,而这也成为设计"漂浮"的儿童纪念馆的强大力量。

为了锚定室内中庭的主题,"记忆展馆"和"宽容展馆"的建筑体量设计就像张开的双臂,拥抱着儿童纪念馆。儿童纪念馆的设计有两个相互关联的寓意:记住大约二百万在种族灭绝大屠杀中被屠杀的儿童,教育我们的孩子人与人之间要学会和平共处。

在室内中庭,博物馆各自不同的功能作为独立的体量来解读。博物馆的永久展厅(记忆与宽容)就展示在L形清水混凝土体量的后面。

宽容与记忆博物馆之旅从最顶层开始。参观者站在悬浮在空中的儿童纪念馆上面,鸟瞰外面真实的自由世界(墨西哥城艺术宫、外交部、华雷斯广场、国家公证档案馆等等),然后准备离开明媚的阳光进入人类历史上最黑暗的时刻。

宽容与记忆展馆分布在博物馆的顶部三层(三层、四层、五层)。从顶层缓缓而下,四层和五层的展厅展览着"记忆"展品,包括与犹太人大屠杀、亚美尼亚、南斯拉夫、卢旺达、危地马拉、柬埔寨和达尔富尔有关的违背人性的种族灭绝和其他罪行。

从"记忆展区"过渡到"宽容展区",参观者要暂时离开展厅到达中庭,进入立面为橄榄枝图案的儿童纪念馆(与荷兰艺术家Jan Hendrix共同设计),里面全部采用自然光线,一串串共20 000滴"眼泪"如瀑布般从空中而下,象征着20 000名受害儿童——每滴眼泪代表着100个消失的灵魂。

走出儿童纪念馆,参观者沿着一段空中楼梯向下,进入一条水晶般透明的走道。一幅由墨西哥艺术家Gustavo Aceves创作的壁画标志着参观者重返三楼的以"宽容"为主题的永久展厅。这一部分由19个展厅组成,展厅设计都富有教育意义,反映了成见、偏见、歧视、仇恨和暴力等主题。空间的设计意在让参观者意识到对话的重要性,建筑师的态度、言语和行为的含义,尊重和包容社会、文化和宗教的差异性的重要性,以及每个人促进和平和防止未来产生的歧视和不宽容的责任。

为了使参观者在参观完后进行最后的反思,建筑师与墨西哥艺术家海伦·埃斯科贝多(2009年艺术与科学国家奖获得者)共同合作,设计了一处隐蔽的、僻静的反思空间。这一空间是个极简抽象派艺术风格的房间,层高很高,一个悬浮的吊顶/平台上下不断移动,给人以压抑和释放的感觉。

因为整个展览不适合12岁以下的儿童,所以建筑师设计了一处专门的凹陷区,通过在工作坊里工作、做游戏和讲故事的方式,向孩子们展示宽容、尊重和多样性的价值。博物馆还包含一个自助餐厅和一个商店。

当参观者即将结束本次博物馆之旅,他最后可以看到一扇窗户,透过窗户可以看到街对面的贝尼托·华雷斯纪念馆。纪念馆主要纪念贝尼托·华雷斯这位伟大的主张自由的墨西哥领导人。人们永远不会忘记他的名言:"个人之间,如同国家之间,尊重他人的权利就是和平。"

Museum of Memory and Tolerance

Museum of Memory and Tolerance integrates the remembrance of genocides provoked by racial discrimination (Memory) and the unforgiving legacy that leaves us with respect of others and co-existence in diversity (Tolerance). This museum provides Mexico a space of study within a democratic and multicultural frame for the development of future generations.

The Museum is constructed using a mixture of reinforced concrete and steel in a seven level structure (three of permanent exhibit + four of complimentary exhibit). It is set on a continuous colonnade of the Plaza Juarez complex, designed by Legorreta + Legorreta Architects on the site of the former "Hotel Alameda" which fell during the 1985 earthquake.

Arditti + RDT Arquitectos designed the museum with the rooted belief that the only hope for humanity lies in the education of future generations. Therefore, the main force behind the conceptual idea of the Museum is sustaining the "floating" Children's Memorial.

In order to anchor this main motif of the interior atrium, the volume that contains Memory and Tolerance is displayed like two open arms embracing the Children's Memorial. This Memorial has two interrelated intentions: remembering approximately two million children who have been exterminated in genocides, and educating our children to foster future coexistence among all of the people.

In the interior atrium, the different functions of the building are read as independent volumes. The Museum's Permanent Exhibits (Memory and Tolerance) are held behind the exposed concrete "L" shaped mass.

The journey through Memory and Tolerance begins on the upper level of the Museum. Standing above the suspended Memorial, the visitor overlooks the reality of the free outside world (The Mexican Palace of Fine Arts, The Secretary of Foreign Affairs, The Plaza Juarez Square, The National Notaries Archive, etc..) and is about to be moved from direct sunlight into some of the darkest episodes of mankind.

Memory and Tolerance are contained on the top three levels of the Museum (5th, 4th and 3rd). Descending from the upper level, Memory is displayed in exhibition halls on the top two floors. Included in these exhibitions are genocides and crimes against humanity relating to the Holocaust, Armenia, Former Yugoslavia, Rwanda, Guatemala, Cambodia and Darfur.

Transitioning from Memory into Tolerance, the visitor is temporarily taken outside to the atrium into the olive skinned Children's Memorial (created in collaboration with the Dutch artist Jan Hendrix) within a naturally lit space, where a cascade of 20,000 "tears" symbolizes the victims – one for each 100 vanished souls.

Exiting the Memorial, the visitor moves down a staircase above the open space and into a crystal walkway. A mural by the Mexican artist Gustavo Aceves marks the re-entry to the permanent exhibition on Tolerance on the third floor. This section is comprised of 19 exhibition halls designed to didactically reflect on topics such as stereotype, prejudice, discrimination, hate and violence. The space is designed to create awareness of the importance of dialogue, the implication of architects' attitudes, words and actions; the importance of respecting and embracing social, cultural and religious differences; and the responsibility of each individual to promote peace and prevent future discrimination and intolerance.

Enabling to make a final reflection, a secluded, quiet introspection space was created in collaboration with the Mexican artist Helen Escobedo (recipient of the National prize of Art and Science 2009). The space is a minimalist room of tall proportion where a suspended ceiling/platform moves constantly downward and upward, with an oppressing and liberating sensation.

Since the overall exhibit is not apt for children under 12, a specific sunken area was conceived where through workshops, games and stories, children are shown the value of tolerance, respect and diversity. The Museum also contains a cafeteria and a store.

As one ends the journey, a final window frames the exterior view across the street towards the Benito Juarez Memorial, where the great Mexican leader who advocated for freedom is remembered. His famous words will never be forgotten: "Among Individuals, as Among Nations, Respect for the Rights of Others Is Peace."

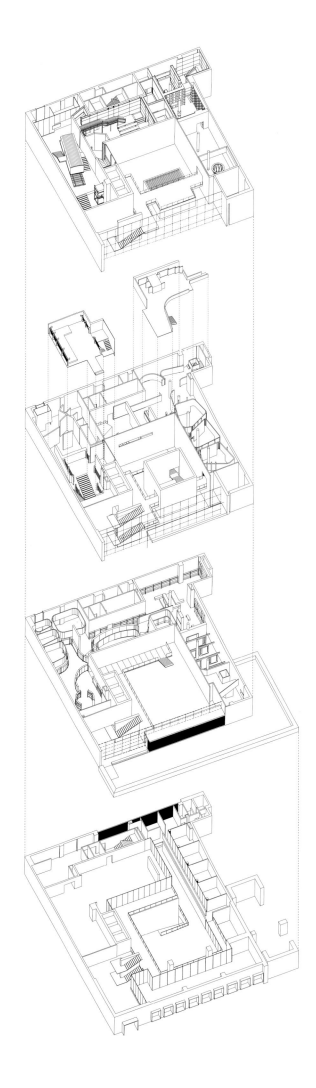

1	垂直流线 vertical circulation
2	咖啡商店 coffee shop
3	办公室 offices
4	临时展馆 temporary exhibitions
5	策展室 curation room
6	教育中心 educative center
7	图书馆和媒体室 library and media room
8	"宽容"主题的永久性展馆 tolerance permanent exhibition
9	露台 terrace
10	"记忆"主题的永久性展馆 memory permanent exhibition
11	儿童纪念馆 children's memorial

五层 fifth floor

四层 fourth floor

二层 second floor

三层 third floor

一层 first floor

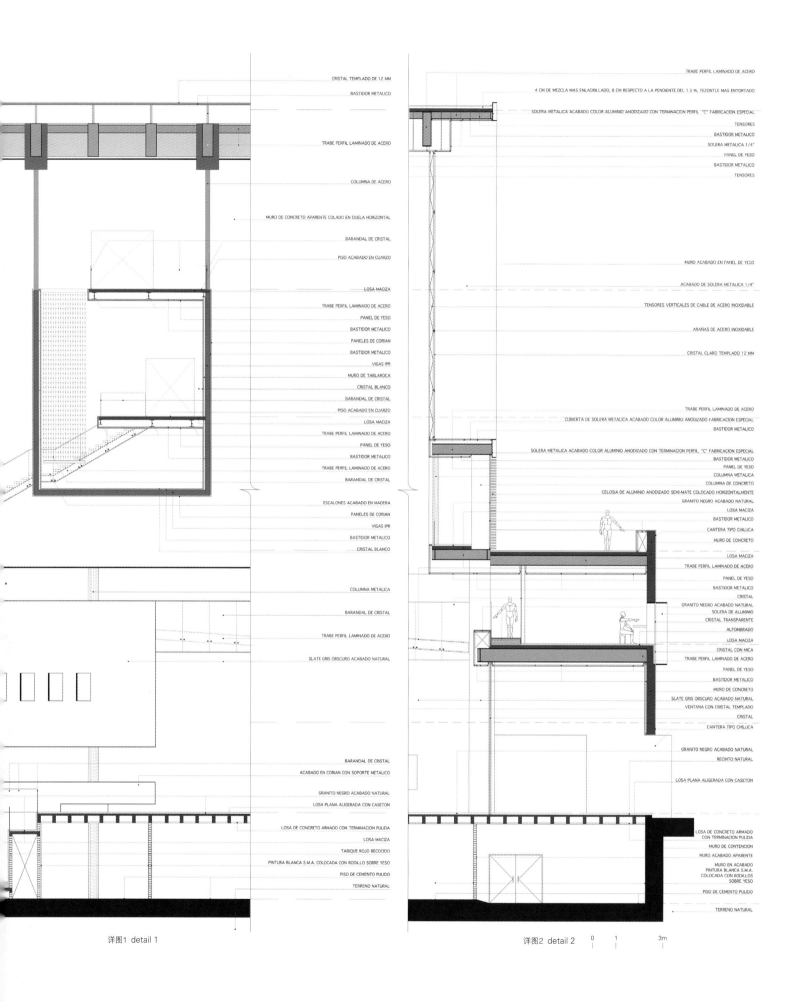

详图1 detail 1　　　　　详图2 detail 2

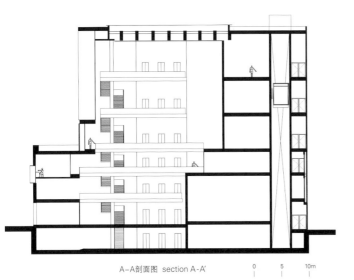

A-A剖面图 section A-A'

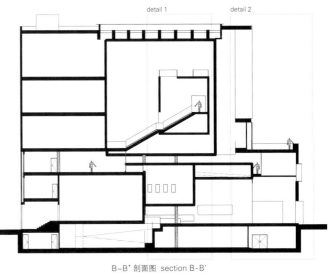

B-B' 剖面图 section B-B'

C-C' 剖面图　section C-C'

D-D' 剖面图　section D-D'

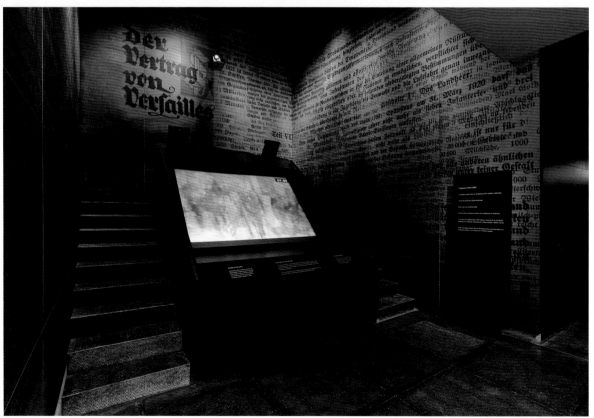

荷兰新国家博物馆

Cruz y Ortiz Arquitectos

2001年,来自西班牙塞维利亚的Cruz y Ortiz建筑师事务所被政府首席建筑师乔·科嫩所领导的委员会选中,负责国家博物馆的改建项目的设计。按照欧洲独立投标过程,Van Hoogevest建筑事务所负责改建项目的修复过程。

Cruz y Ortiz建筑师事务所将一栋19世纪的建筑改造成了一座21世纪宽敞明亮的博物馆。所有原来建筑的增建部分,如低矮的天花板吊顶和半层阁楼建筑,都被拆除,以确保其再次成为一个连贯的整体。Cruz y Ortiz建筑事务所设计了一个引人入胜的新入口,称为中庭(Atrium),以及一座新的亚洲馆和一座用作服务入口的新建筑。参观者可以享受先进的设施,包括现代的咖啡馆、商店、礼堂和首次对外开放的重建的图书馆。设计师们还设计了工作室大楼,该建筑于2007年开放,是负责国家博物馆修复项目的各个工作室的办公所在地。从文物收藏保护和气候控制措施方面来说,此建筑十分符合当今的最新需求。

21世纪Cuypers设计的建筑

原来墙壁和天花板上的纪念装饰还进行了修复,应用在一些大名鼎鼎的展厅,包括荣誉画廊、大礼堂、夜巡画廊和楼梯间。在Van Hoogevest建筑事务所的要求下,已从大礼堂消失的水磨石地面得到了完全恢复。Cuypers建筑的设计特点在图书馆得到了最好的保护,原设计和装饰品得到最完好的维护和修复。

中庭

Cruz y Ortiz建筑师事务所把原来的内部庭院改造成一处众所周知的新入口区域,称为中庭。中庭装有大型的玻璃屋顶和白色抛光葡萄牙石材地板。地板可反射自然光,使得宽敞的庭院富有清新明亮之感。周围博物馆建筑给人温暖之感的砖立面俯瞰着庭院,窗户和壁龛点缀其间。

一条中央通道可通往中庭。这里,原来的砖墙被玻璃幕墙取代,路人可以由此欣赏庭院的美景。

新的中庭入口对所有参观者开放,无票也可参观。入口区域设有咖啡厅、商店、服务台、售票厅和衣帽间。

亚洲馆

独立的亚洲馆位于国家博物馆南部的花园内,四面环水。这一形状不规则的两层建筑的外墙采用白色葡萄牙石材和玻璃幕墙来建造,与主楼的红砖墙形成鲜明对照。其特色是设有许多倾斜的墙体,因此产生非同寻常的视野。亚洲馆通过地下通道与博物馆主楼相连接。

亚洲馆用于展出公元前2000年到公元2000年间来自中国、日本、印度尼西亚、印度、越南和泰国的文物和艺术作品。该馆藏有丰富的亚洲艺术品,在其485m²的空间里和谐地展出了大约350件展品。

工作室大楼

工作室大楼是国家博物馆修建项目的一部分,由政府建筑署开发,受国家博物馆和教育、文化和科学部委托,于2007年开放,也是Cruz y Ortiz最先完成的部分。

工作室大楼专为荷兰文化遗产保护和管理而建造,成为进行修复和保护、科学实践、研究和教育最先进的中心。

该建筑容纳了国家博物馆、阿姆斯特丹大学以及荷兰文化遗产局所有与修复工作相关的部门。

工作室大楼表面积超过9000m²,将Pierre Cuypers设计的原建筑和新建筑融为一体。原建筑曾被用作安全研究所,一直是国家博物馆不可或缺的组成部分,功能性是工作室大楼设计中考虑的重中之重。其特别的"Z字形"屋顶结构、玻璃装饰的北侧立面和侧壁上有一定角度的三角形窗户都可以确保只有北侧的光可以照射进室内,避免正在修复的艺术品受到阳光直射。所有的工作室、走廊、门和电梯都比普通的要更高更宽,以便大型艺术品的搬运。

服务入口和能源环

在花园对面,Cuypers别墅和拉丝学校(绘画学校)之间,Cruz y Ortiz设计了另一栋新的小建筑——新的服务入口,可由此通过地下通道访问博物馆。这栋建筑也是"能源环"的入口。"能源环"是一条新的近500m长、环绕整座建筑的隧道,里面容纳了所有的技术设施:气候控制系统、电子设备和安全设施,对整个博物馆的运作至关重要。

The New Rijksmuseum

In 2001, Spanish architects, Cruz y Ortiz from Seville, were chosen by a committee chaired by chief government architect Jo Coenen to create a new design for the Rijksmuseum. In a separate European tender, Van Hoogevest Architecten was selected for the restoration. Architects Cruz y Ortiz have transformed the 19th-century building into a light and open 21st-century museum. All later additions to the original building, such as the lowered ceilings and half-stories, have been removed to ensure that it is once again a coherent whole. Cruz y Ortiz have created a spectacular new entrance, called the Atrium, as well as a new Asian pavilion and a new building that serves as the service entrance. Visitors can enjoy state-of-the-art facilities, including a modern cafe, a shop, an auditorium and, for the first time, the restored library. The architects also designed the Atelier Building, where the Rijksmuseum's restoration studios are housed, which opened in 2007. The building satisfies the latest requirements in terms of preservation of the collection and climate control measures.

Cuypers for the 21st Century

The original monumental ornaments that decorated the walls and ceilings have been restored in prominent rooms including the Gallery of Honor, the Great Hall, the Night Watch Gallery and the stairwells. The terrazzo floor that had disappeared from the Great Hall also has been fully restored at Van Hoogevest Architecten's behest. Cuypers' hallmark is best preserved in the library, where the original design and ornaments have been beautifully maintained and renovated.

1 主楼	1. main building
2 工作室大楼	2. atelier building
3 亚洲馆	3. Asian pavilion
4 入口大楼	4. entrance building
5 Philips翼楼	5. Philips wing
6 别墅	6. villa
7 绘画学校	7. drawing school

Atrium

Architects Cruz y Ortiz have turned the former inner courtyards into an impressive new entrance area, known as the Atrium. The space features large glass-covered roofs and pale polished Portuguese stone floors that reflect the natural light, making the voluminous courtyard spaces feel bright and airy. Overlooking the courtyards are the warm brick facades of the surrounding museum buildings, interspersed with windows and niches.

The Atrium can be accessed from the central passageway, where the original brick walls have been replaced by glass through which passers-by can admire the view of the courtyards.

The new Atrium entrance is open to all visitors, including those without an admission ticket. This area includes the cafe, the shop, the information desk, ticket sales and the cloakroom.

The Asian Pavilion

This free-standing pavilion is surrounded by water and situated in the gardens to the south of the Rijksmuseum. The irregular-shaped, two-story structure, with its walls of pale Portuguese stone and glass, stands out against the red brick walls of the main building. It is characterized by many oblique walls and unusual sightliness. The pavilion is linked to the main building via a ground-level passageway.

The Asian Pavilion has been created to showcase objects and works of art from China, Japan, Indonesia, India, Vietnam and Thailand, dating from 2000 B.C. to 2000 A.D. The museum's rich collection of Asian art is harmoniously presented in the 485 square-meter space, with approximately 350 objects on display.

Atelier Building

Developed by the Government Buildings Agency and commissioned by the Rijksmuseum and the Ministry of Education, Culture and Science (OCW), the Atelier Building opened in 2007 as the first structure that Cruz y Ortiz completed as part of the Rijksmuseum renovation.

It is intended for the preservation and management of Dutch cultural heritage, and is home to a state-of-the-art center for restoration and conservation, scientific practice, research and education. It accommodates all restoration departments of the Rijksmuseum, the University of Amsterdam and the Netherlands Cultural Heritage Agency.

Covering a surface area of more than 9,000 square-meters, the Atelier Building merges a new structure with an existing building designed by Pierre Cuypers, known as the Safety Institute, which has always been part of the Rijksmuseum. Functionality was paramount in the building's design. The unusual "zigzagging" roof structure, the glazed northern elevation, and the angled, triangular windows in the side wall ensure that only northern light is admitted, protecting the works that are being restored from direct sunlight. All the studios, hallways, doors and lifts are higher and wider than usual, facilitating the transportation of large works of art.

Service Entrance and Energy Ring

On the opposite side of the gardens, between the Cuypers Villa and the Teekenschool (Drawing School), Cruz y Ortiz have placed another small building – the new service entrance, offering access to the museum via an underground passageway. This building is also the entrance to the "Energy Ring", a facility crucial to the museum's operation. The new, nearly 500m-long tunnel around the building houses all the technical facilities: the climate control system, electronics and security.

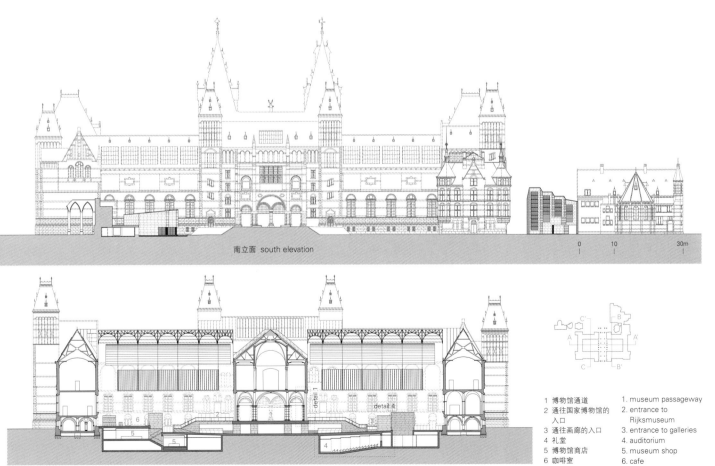

南立面 south elevation

A-A' 剖面图 section A-A'

1. 博物馆通道 / museum passageway
2. 通往国家博物馆的入口 / entrance to Rijksmuseum
3. 通往画廊的入口 / entrance to galleries
4. 礼堂 / auditorium
5. 博物馆商店 / museum shop
6. 咖啡室 / cafe

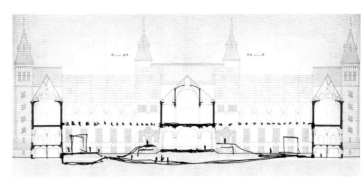

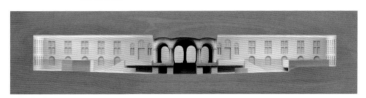

项目名称：The New Rijksmuseum
地点：Amsterdam, Netherlands
首席建筑师：Antonio Cruz, Antonio Ortiz
项目建筑师：Muriel Huisman, Thomas Offermans
项目团队：Oscar García de la Cámara, Tirma Reventós, Alicia López, Marije ter Steege, Juan Luis Mayén, Jan Kolle, Victoria Bernícola, Clara Hernández, Ana Vila, Joaquin Pérez, Marta Pelegrín, Sara Gutiérrez, Iko Mennenga, Lourdes Gutierrez, Carlos Arévalo
修复建筑师：Van Hoogevest Architecten
本地建筑师：ADP Architecten
花园设计：Copijn Tuin-en Landschapsarchitecten
结构和土木工程师：Arcadis
机械工程师：DGMR, Royal Haskoning
消防安全：DGMR
建筑物理：DGMR
照明咨询：Arup
主承包商：JP van Eesteren
分包商：Koninklijke Woudenberg, Bam Civiel, Homij, Kuipers, Moehringer, Unica
场地监督：BRINK
总建筑面积：30,000m²
总展览面积：12,000m²
中庭面积：2,250 m²
亚洲馆面积：670m²
花园/室外博物馆面积：14,500m²
造价：EUR 375,000,000
竣工时间：2013
摄影师：©Iwan Baan (courtesy of the architect) - p.84, p.96middle, p.96bottom-left, p.97
©Pedro Pegenaute (courtesy of the architect) - p.88~89, p.91top, p.92~93
©Pedro Pegenaute - p.91bottom, p.94 (except as noted)

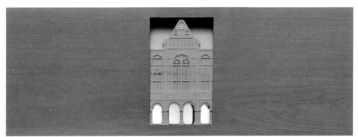

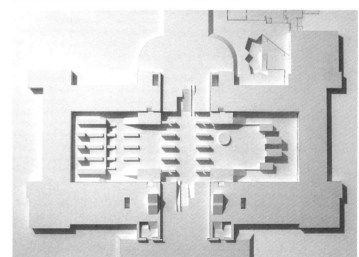

B-B' 剖面图 section B-B'

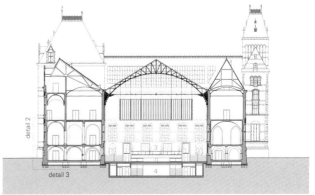

C-C' 剖面图 section C-C'

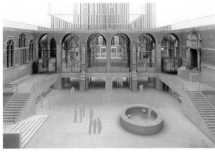

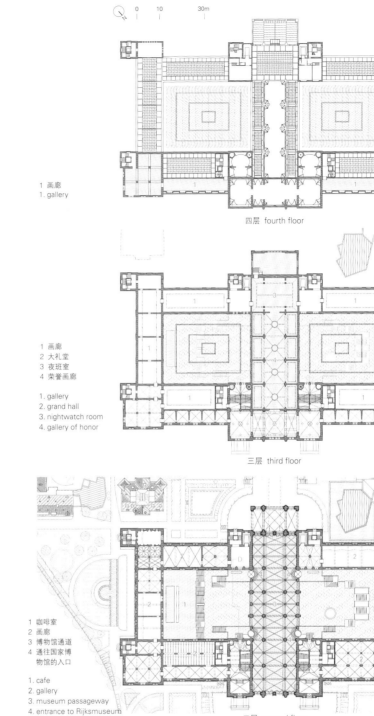

1 画廊
1. gallery

四层 fourth floor

1 画廊
2 大礼堂
3 夜班室
4 荣誉画廊

1. gallery
2. grand hall
3. nightwatch room
4. gallery of honor

三层 third floor

1 咖啡室
2 画廊
3 博物馆通道
4 通往国家博物馆的入口

1. cafe
2. gallery
3. museum passageway
4. entrance to Rijksmuseum

二层 second floor

1 信息台
2 售票处
3 衣帽间
4 通往画廊的入口
5 博物馆商店
6 卫生间
7 画廊
8 亚洲馆
9 Philips翼楼
10 入口大楼

1. information desk
2. ticket desk
3. cloak room
4. entrance to gallery
5. museum shop
6. toilets
7. gallery
8. Asian pavilion
9. Philips wing
10. entrance building

地下一层 first floor below ground

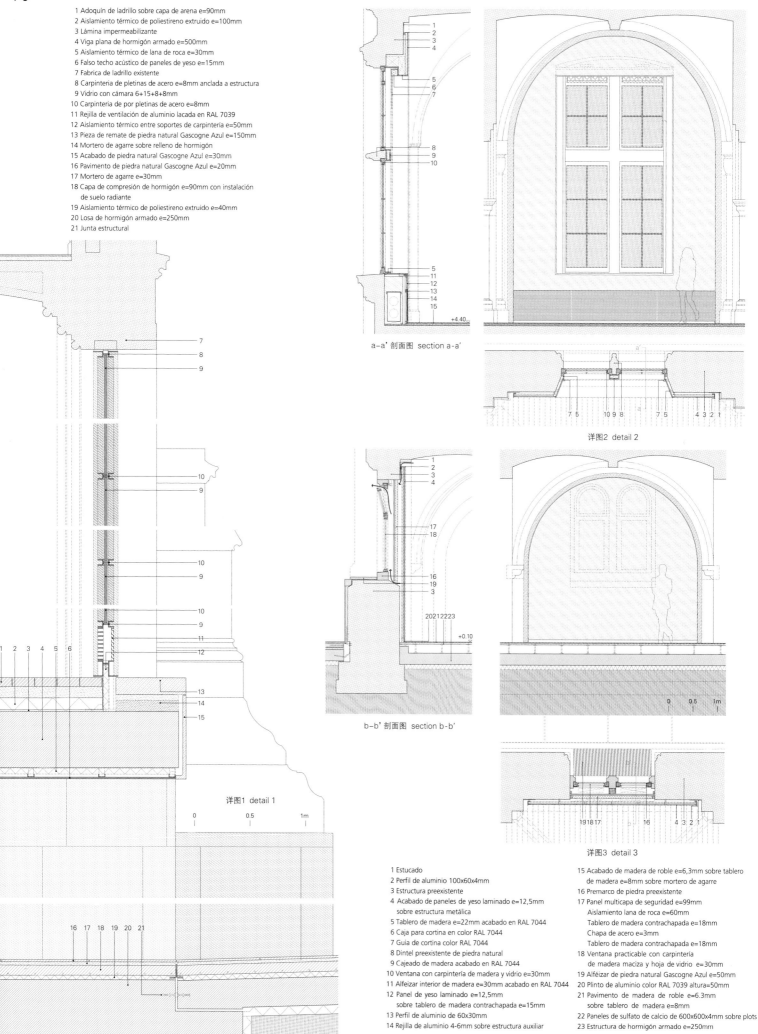

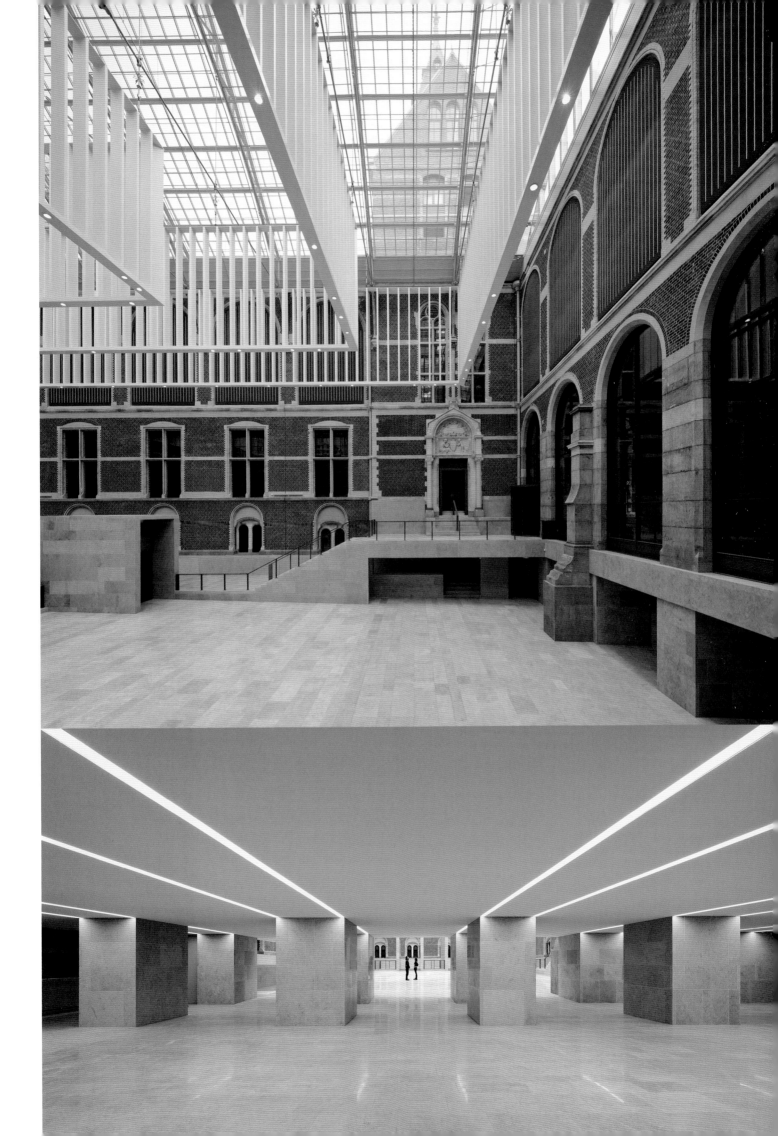

1 Sujeción a cercha. Estructura existente
2 Perfil en U deformable
  (asientos diferenciales entre estructura existente y estructura nueva)
3 Varilla roscada
4 Perfil de acero 100x245x10 mm
5 Perfil de acero L 150.100.10
6 Cableado iluminación
7 Pieza de madera DM e=18 mm pintada en RAL 9010
8 Panel acústico de madera de 84x336 mm pintado en RAL 9010
9 Estructura interior de madera maciza de 40x60 mm
10 Aislamiento de lana mineral e=60 mm
11 Iluminación LED
12 Iluminación puntual

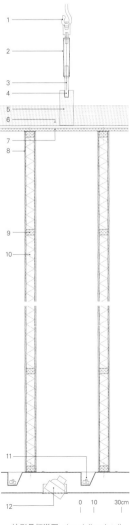

枝形吊灯详图 chandelier detail

D-D' 剖面图 section D-D'

1 Pavimento de piedra natural Gascogne Azul e=20mm sobre mortero de agarre
2 Capa de compresión de hormigón e=100mm con instalación de suelo radiante
3 Aislamiento de poliestireno extruido e=40mm
4 Estructura de hormigón armado e=250mm
5 Acabado de piedra natural Gascogne Azul e=30mm
6 Cajeado de aluminio e=3mm para iluminación general
7 Rejilla metálica 4-6mm sobre estructura auxiliar
8 Pieza de remate de piedra natural Gascogne Azul
9 Capa de compresión de hormigón armado e=50mm
10 Aislamiento de poliestireno extruido e=100mm
11 Pantallas de contención de acero de 11m de longitud
12 Perfil metálico interrupción de pavimento en junta
13 Junta estructural
14 Pavimento de piedra natural belga sobre mortero de agarre
15 Cimentación estructura existente
16 Pavimento de madera de roble e=6.3mm sobre tablero de madera e=8mm
17 Paneles de sulfato de calcio de 600x600x4mm sobre plots
18 Pórtico de madera h=2400mm (ver detalle específico)
19 Panel de yeso laminado sobre estructura auxiliar
20 Acabado de piedra natural Gascogne Azul e=30mm anclado a estructura de hormigón
21 Cajeado de chapa plegada de aluminio e=3mm para iluminación general
22 Dintel de madera DM sobre estructura auxiliar
23 Viga de hormigón armado 400x700mm
24 Barandilla de pletinas de acero e=8mm con vidrio
25 Falso techo de panel de yeso laminado e=15mm
26 Losa hormigón armado e=180mm
27 Viga de hormigón armado 250x900mm empotrada en estructura preexistente
28 Estructura preexistente
29 Relleno de mortero
30 Pletina de acero e=3mm
31 Carpintería de acero RF-30min
32 Doble hoja de vidrio con cámara e=6+15+8mm

详图4 detail 4

1 Pavimento de madera de roble e=6.3mm sobre tablero de madera e=8mm
2 Paneles de sulfato de calcio de 600x600x4mm sobre plots
3 Losa de hormigón armado e=300mm
4 Aislamiento de poliestireno extruido e=65mm
5 Capa de arena e=100mm
6 Hormigón en masa e=800mm
7 Pieza de remate de madera maciza de roble e=27mm
8 Subestructura metálica construcción graderío
9 Muro de hormigón armado e=250mm
10 Aislamiento de poliestireno extruido e=100mm
11 Paneles de madera e=12mm lacada en RAL 9010
12 Pantalla para proyecciones
13 Cortina corredera
14 Estructura auxiliar para iluminación de eventos
15 Paneles de yeso laminado sobre estructura auxiliar
16 Cajeado de aluminio e=3mm para iluminación general
17 Junta estructural
18 Pavimento de piedra natural Gascogne Azul e=20mm
19 Capa de compresión e=90mm con suelo radiante
20 Aislamiento de poliestireno extruido e=40mm
21 Losa de hormigón armado e=500mm
22 Murete de ladrillo
23 Techo acústico de madera e=16mm
24 Revestimiento de paneles acústicos de madera
25 Carpintería de madera maciza lacada en RAL 9010
26 Techo de paneles de madera e=16mm
27 Panel sandwich:
   Aislamiento interior de lana de roca e=72mm
   Paneles de fibrocemento e=15mm
   Paneles de yeso laminado e=9mm
28 Falso techo de paneles de madera e=30mm
29 Rejilla metálica de ventilación
30 Acabado de piedra natural Gascogne Azul e=30mm
31 Estructura de hormigón armado e=250mm

博物馆的变迁 Multi-Museum

国立Machado de Castro博物馆
Gonçalo Byrne Arquitectos

二千年前，在科英布拉山的中央，由两个拱形层支撑的一个矩形平台——一个罗马式拱形画廊建造起来，以容纳罗马Aeminiun城市的中央论坛。在时间的长河中，论坛消失了，但这个画廊因其具有完美的构造，所以几乎完好无损地保留了下来。在西方欧洲建筑悠久的历史上，它就像是一张表格，而在这份表格中，这些按照时间顺序排列的建筑碎片会被添加或混合在一起，有时则会减掉或切割掉一部分建筑。

这些伊斯兰的、罗马式的、哥特式的、文艺复兴式的、巴洛克式的、浪漫复兴现代派的建筑碎片也反映出其在不同时期的不同用途，从中央罗马论坛的用途不一，到用作仓库、街铺、附加建筑、马厩等等。从12世纪到20世纪早期，这里变成了基督教主教的住处。1910年，这里的所有街区最后被改建成一座城市博物馆，主要展览杰出的16世纪文艺复兴时期的作品。在20世纪60年代，一座珍贵的、文艺复兴时期的、名为Jean de Rouen的教堂从原址拆除，被重建成一座"格格不入"的博物馆，为了创造恰当的且备受保护的展览氛围，建筑师对其设计做了很大的干预。

在这样一个无论从时间还是外形上来说都非常复杂的地方重新设计一座新博物馆，这一任务既艰巨又令人着迷。因此，无论从参观博物馆内部来看，还是从周围景观（参观者从沉浸在对画廊的沉思冥想中到感知整个城市不同层次的宽阔视野，可以漫步而下，直达河谷）的不同洞口来说，在通向这一刻着时间印记的、传承历史的建筑过程中，漫步成为感受生活的实质性体验。

有两块相邻的空地，一块可用于修建行政办公楼，另一块离博物馆较近的可用来修建技术服务区、基础设施和安装垂直电梯，这样参观者可以不必侵扰这一具有历史意义的紧凑的建筑肌理就能到达新博物馆的每一层。

这座新体量更加突显了之前的罗马论坛的原有平台，其向外延伸出来的全景露台形成一种非常壮观的双重视觉联系，既可以饱览这一历史名城的优美风光，又可以近观Fillipo Terzi文艺复兴时期的柱廊和其围成的令人震撼的庭院。

主入口保留了原来进入前主教宫殿的大门和宫殿中央的庭院，庭院是再现罗马论坛必不可少的一部分。在这里，人们可以提前感受到本次参观所带来的感官体验。参观可以从隐藏的地下空间开始，人们或者通过Terzi柱廊到达地下空间的边界来感受，或者进入不同交通流线两侧的空间，或者直接过桥到达博物馆餐馆和餐馆露台。

博物馆的主体量非常精确地控制在最初的罗马论坛周界之内，而画廊的外立面在与城市的交融中得到了很好的增强效果。公共楼梯把画廊与相邻的新建筑分离开来，既让参观者在新博物馆真实感受古老建筑最初存在的样子，也让他们真实感受到此刻正位于科英布拉市的最中心点。参观者的当代体验说明所有的手工艺品都是周围城市的聚光镜，人们在断断续续的或较完整的感知中都可以感受到其更多的相似之处、可类比之处，以及对比之处。

Machado de Castro National Museum

Two thousand years ago, up in the central hill of Coimbra, a rectangular platform supported by two vaulted layers – a Roman criptoporticum – was built to contain the central forum of the Roman city of "Aeminiun". This forum disappeared in time but the Criptoporticum remained almost intact in its powerful tectonics, acting like a sort of table where sequential fragments add, mix, sometimes subtract, cut off architectonic pieces of the long history of western European architecture.

These fragments of Islamic, Romanesque, Gothic, Renaissance, Baroque, Romantic Revival Modernity also reflect different uses along time, from the central mixed use of the Roman forum, to stocking warehouse, street shops, attached housing, horse stables, etc.. In Christianity it became the Episcopal residence from the twelfth century up to the early twentieth. In 1910 all the blocks

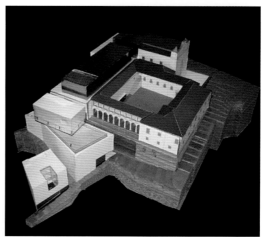

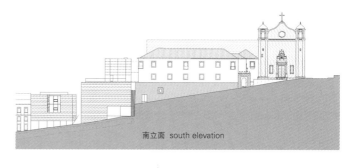

南立面 south elevation

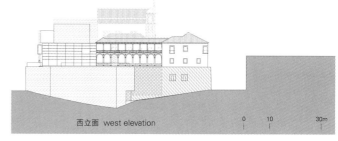

西立面 west elevation

were finally converted into a city museum, displaying mainly a remarkable renaissance collection of XVI century. In the sixties a precious "Jean de Rouen" renaissance chapel was dismantled from its original site and reassembled as an "Out of Scale" museum object, allowing a more substantial intervention to create an appropriate and protected exhibition atmosphere.

To recreate a new museum out of such time and shape complexity is a very demanding but fascinating task. A promenade through the fragmentary time-printed condition of the inherited architecture is an essential experience to live, both from the internal point of view, and also from the differentiate openings to the surroundings – from total introspection of criptoporticum to wide open perception of the long-term city layers, cascading down to the river valley.

Two adjacent empty plots are available first for the administration offices and the closer one to technical services, infrastructure and vertical accessibility so that one can reach every level of the new museum without intruding on the historic compact fabric. This new volume underlines the original platform level of the ex-forum extending a panoramic terrace which creates an intense dual visual relationship with either the historic city, or the close presence of the Fillipo Terzi renaissance portico and its powerful framing of the courtyard.

The main entrance keeps the original gate to the ex-bishop's palace and its central courtyard as an essential void where the potential of the ex-forum can be recreated, a space where people can anticipate all the sensitive experience of the visit, starting with the hidden underground void structure either coming to its edge through the Terzi portico or entering the side spaces of different circuits, or simply bridging into the museum restaurant and its terrace.

This main volume is very precisely limited to the original perimeter of the forum and the external facade of the Criptoporticum is strongly enhanced in its urban intermediation, separated from the adjacent new building by a public staircase that makes real the experience of its germinal presence in the new museum as well as the foundational center point of the city of Coimbra. The contemporary experience of the visit reveals all the artifacts as a sort of condenser of the surrounding city where more similarities, analogies, contrasting perspectives can be realized either in fragmentary or more unitarian perception.

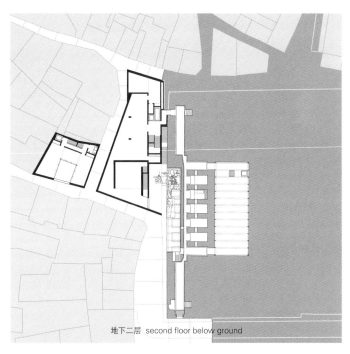
地下二层 second floor below ground

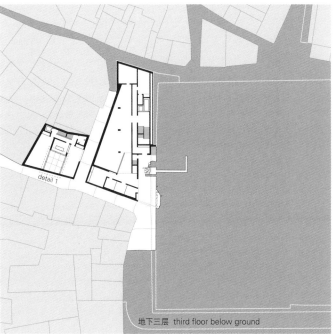
detail 1
地下三层 third floor below ground

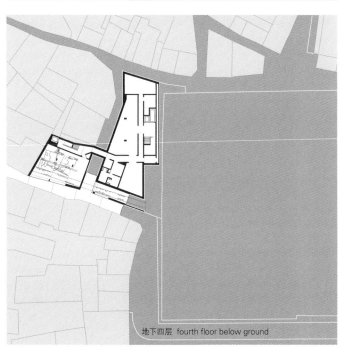
地下四层 fourth floor below ground

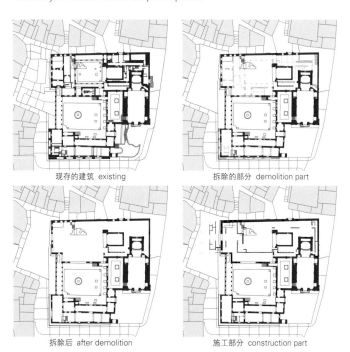
现存的建筑 existing　　拆除的部分 demolition part
拆除后 after demolition　　施工部分 construction part

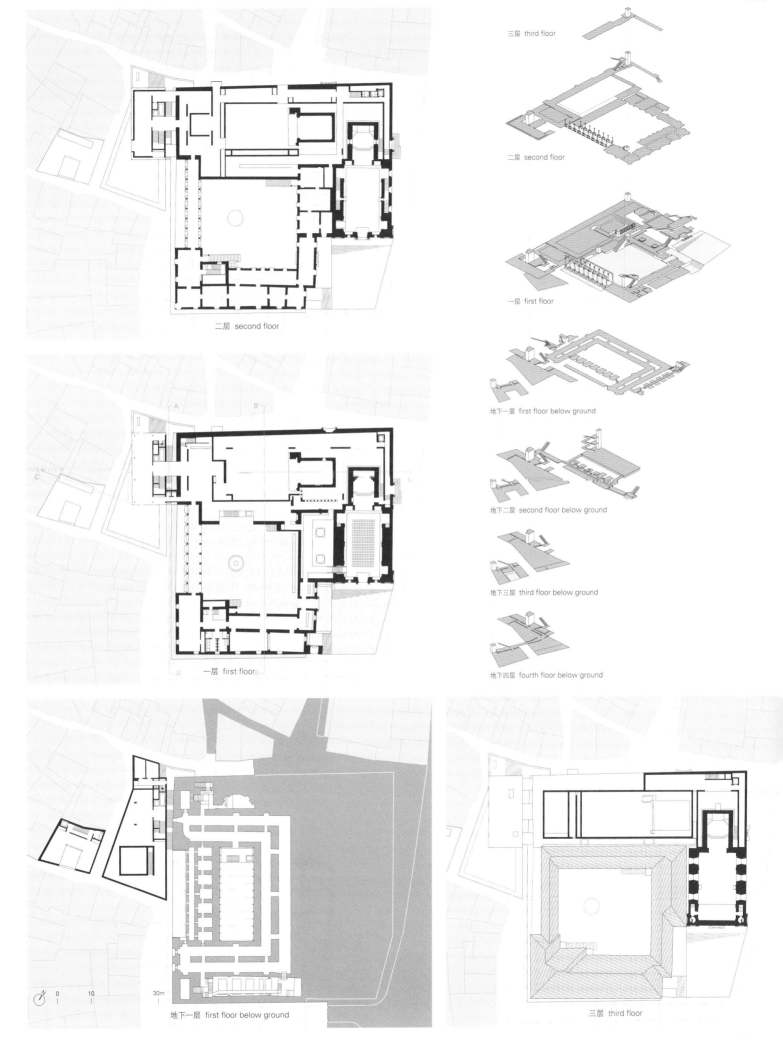

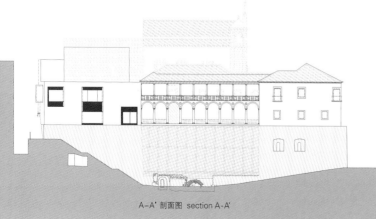

A-A'剖面图 section A-A'

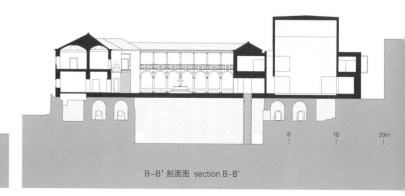

B-B'剖面图 section B-B'

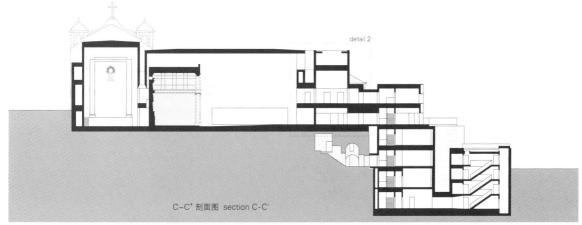

C-C' 剖面图 section C-C'

详图1_百叶窗的操作
detail 1_operation of the louver

a-a' 剖面图
section a-a'

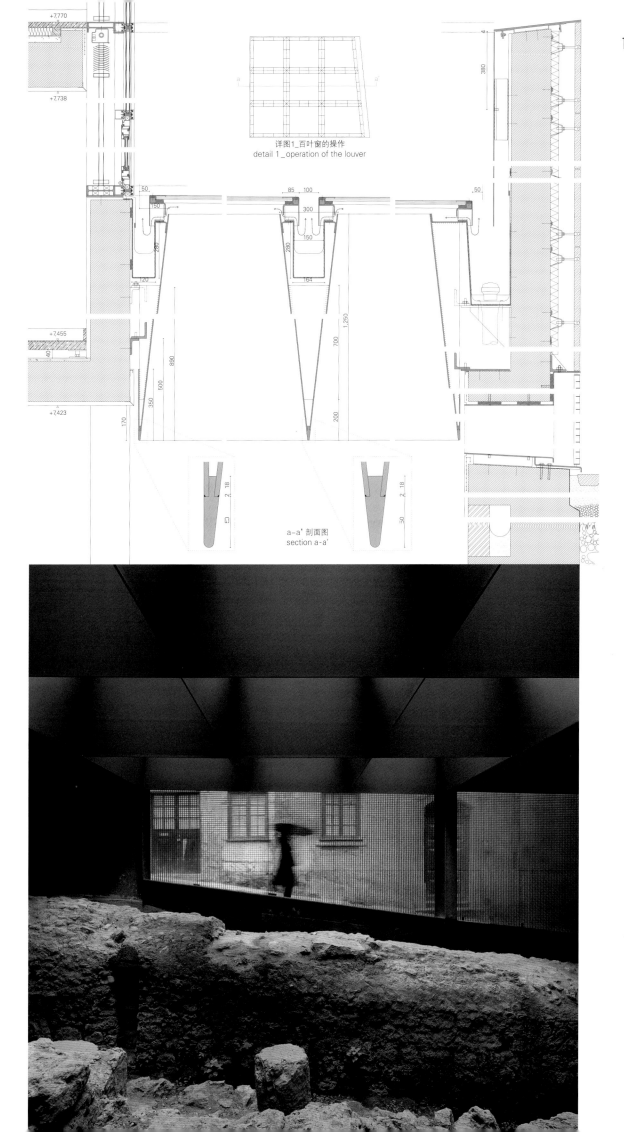

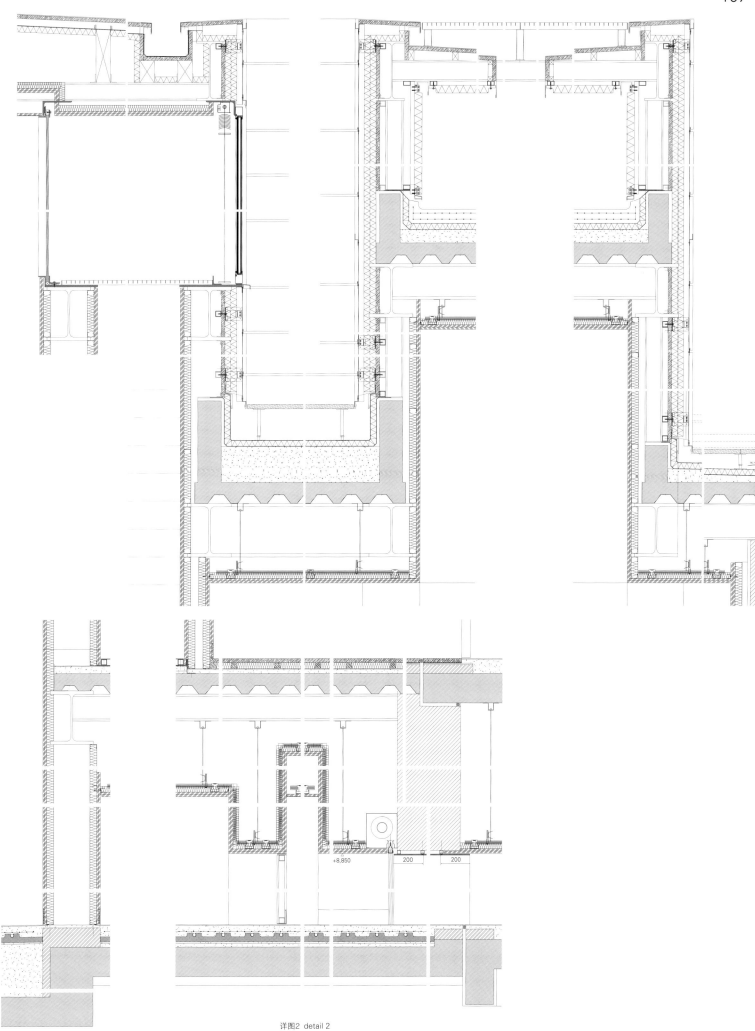

详图2 detail 2

项目名称：Machado de Castro National Museum
地点：Coimbra, Portugal
建筑师：Gonçalo Byrne
项目管理：José Barra, Nuno Marques, Maria João Gamito
合作商：Alexandre Berardo, Ana Conceição, Ana Filipa Santos, Ana Natividade, Catarina Sousa, Gonçalo Lopes, Gustavo Abreu, Joana Sarmento, João Gois, José Laranjeira, José Martins Pereira, Leonor Raposo, Mafalda Rebelo, Margarida Silveira Machado, Maria João Costa, Marta Oliveira Dias, Miguel Lira Fernandes, Nuno Birne, Nuno Fidéles, Patrícia Barbas, Patricia Caldeira, Pedro Neves, Rita Freitas, Rodrigo Germano, Rolf Heinemann, Telmo Cruz, G/F Arquitectos Associados Lda,
结构顾问：BETAR, Estudos e Projectos de Estabilidade Lda.
机械顾问：José Galvão Teles Engenheiros Lda
电气顾问：JOULE Projectos Estudos e Coordenação Lda.
照明顾问：Raul Serafim
博物馆技术顾问：Dra Adilia Alarcão, Dra Ana Alcoforado
甲方：Instituto dos Museus e da Conservação
施工面积：13,130m²
施工时间：2006—2012
摄影师：©Duccio Malagamba

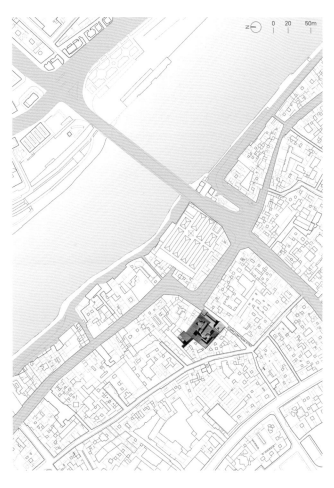

特里亚纳陶瓷博物馆

AF6 Arquitectos

该项目在一个古老的陶器厂建筑群上加以开发，是一个陶瓷展览中心，也是解读特里亚纳街区不同旅游路线以及解读举办圣安娜陶瓷厂的商业和生产活动的各个区域的中心。

从当代的角度来看，该项目是锦上添花，为这个古老的建筑群的共存进行增值。从外部来看，这个建筑群呈现出不同的外表，充分向人们展现了特里亚纳固有的文化。建筑群由两个相互连通的地块组成，三座不同的建筑彼此紧挨着，面向大街的外立面也各不相同。

第一座建筑是特里亚纳陶瓷博物馆入口所在的位置，面对的是圣安娜陶瓷厂独特的广告瓷砖，位于楼群的一角，鸟瞰Altozano广场。Altozano广场是特里亚纳桥的起点。第二座建筑的高度稍低一些，外观看起来有点儿工业化，更加简朴无华。最后，第三座建筑曾是一栋三层楼房，在临街一侧有独立的入口。

　　特里亚纳陶瓷博物馆完全渗入到这一街区建筑群的肌理之中，创造了丰富的城市内部景观。新空间无论从高度还是形状都与现有建筑相契合。此建筑并不旨在成为视觉的焦点，从而改变特里亚纳建筑给人的整体印象。建筑没有外立面，其用意在于参观者进入其中，自己去发现这样的一份礼物。在一楼，参观者可以沿着陶瓷窑炉间的一条连续的通道来进行参观。入口设在San Jorge大街。一楼还有商店、新工作坊和旧陶器厂。不久，新展会将在旧陶器厂举行。

　　博物馆通过将不同的陶瓷元素置于其原始情境中的方式来向参观者讲述陶瓷的生产流程。参观就像是一次迷宫之旅，在窑炉和古老的工厂内穿行，向参观者讲述着特里亚纳传统陶瓷生产是如何进行的。为了达到这个目标，设计中采用了考古方法，不去擦除任何时间和空间的痕迹(烟尘、混乱无序、砖块、木屑、灰尘)，这些元素也成为这一文化遗产的组成部分。

　　新展区位于二层，是一处陶瓷碎片包围的悬浮空间。它是一座独立的建筑，与一层空间的设计原理截然相反。

　　而下层的项目则被设计为一座连接楼层的迷宫。在另一侧，上层的结构与一层分离开来，成为独立的部分，并根据公用隔墙设置了灵活的矩形交通流线。

　　由此产生的外形与每一组窑炉(假设这些窑炉能够重新启动)的几何形状相匹配。在参观这个展览期间，空间先进行拓展，然后又缩小。用于展览的画廊位于更宽敞的区域，而人们在狭窄的区域能够对旧工厂进行了解。之后人们便看见用于展览不同时代陶瓷制品的永久性展览空间：中世纪的、文艺复兴时期的、巴洛克风格盛行时期的，以及19世纪和20世纪陶瓷制品展览空间。

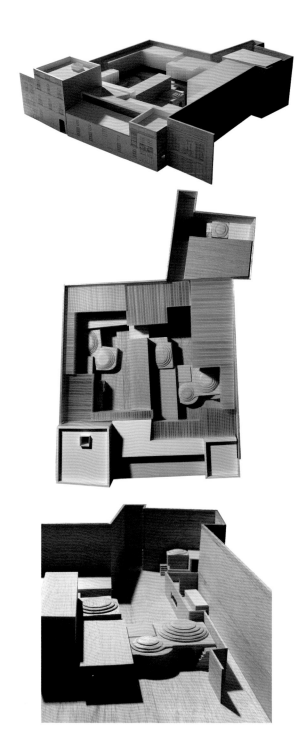

之后，一座带有砖墙的二层老建筑出现在场地的中央。这里是陶瓷画家过去工作的地方，而现在该地则用做临时展厅。在慢慢靠近外面的地方，有一处区域名为"Aquí Triana"，它是为人们解释参观路线的中心。在二层，立面的室内包装强化了积累过程（这座建筑即采用这一原理来建造）的这一理念。镀锌的钢质低层结构可被看做是一个大型货架，为四种不同型号的、凌乱堆放的空心瓷砖构件提供支撑。这一措施是依据太阳方位来避免阳光照射，同时人们还可以从不同的角度来参观陶瓷厂。

Triana Ceramic Museum

The project develops on an old pottery complex, an exhibition center of ceramics, is an interpretative center on different tourist routes in the quarter of Triana, as well as different areas for the commercial and productive activities of Santa Ana Pottery Factory.

The project is proposed as an extra process seen from a contemporary point of view, creating an added value with this complex coexistence. From the outside, the complex shows a heterogeneous image that tells people a story inherently related to the culture of Triana. It consists of two interconnected plots where there are three different buildings attached to each other with different facades onto the street.

The first one provides access to the Triana Ceramic Museum and is faced with unique advertising tiles of Ceramica Santa Ana. It constitutes the corner that overlooks Altozano Square, from which the Triana Bridge starts. The second one, lower in height, has an industrial and more austere appearance and finally, the third one which used to be a three-story building, has an independent entrance from the street.

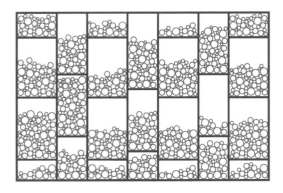

 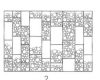  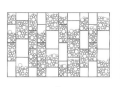 

1 2 3 4 5 6 7 8

9

10
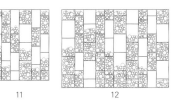
11 12

13
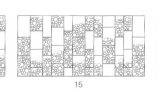
14 15

16

Triana Ceramic Museum seeps into the quarter complex plot tissue, creating a rich urban inner landscape. The new spaces adapt their heights and shapes to the existing buildings. The aim of the project is not to be a visual reference to alter the architectural profile of Triana . There is no facade. The complex is meant to be a gift to be discovered once one gets into it. On the first floor, the building is organized to be visited on a continuous walk among the pottery kilns. The entrance is through San Jorge St.. The ground floor is shared by the shop, the new workshop and the old pottery factory where the new exhibitions will take place.

A narrative of the ceramic production process is carried out by the use of different pottery elements inserted in its original context. It proposes a labyrinthine journey among kilns and old factory spaces which communicates with the visitor about how the traditional ceramic production in Triana was performed. In order to achieve this, an archaeological method is followed without erasing the time and space traces (smoke, disorder, bricks, wood, ashes), these elements constitute a part of this heritage.

The new exhibition area appears on the second floor through a suspended space surrounded by ceramic pieces. It is a separate building with an opposed logic to the one on the first floor.

The project downstairs is meant to be a labyrinth linked to the floor. On the other side, the construction upstairs is physically separated from the first floor with an independent structure configuring a neat rectangular route which relies on the party walls.

The resulting shape adapts to the geometry of each one of the kilns group which could hypothetically start working again. During the tour through the exhibition, space expands and then shrinks. The exhibition galleries are located in the wider areas; from the narrow areas one can have a perspective of the old factory. Then there are spaces for the permanent exhibition of ceramics from different times: Medieval, Renaissance, Baroque, Nineteenth and Twentieth Century.

Then an old two-story building with brick walls emerges in the center of the plot. That is where the ceramic painters used to work and it is now meant to be used for temporary exhibitions. A little further towards the outside, there is an area called "Aquí Triana", created as an interpretive center for tourist routes. On the second floor, the interior wrapper of the facades reinforces the concept of accumulation process with which the project is built. A galvanized steel sub-structure thought as large shelves provides support for an apparently messy pile of hollow ceramic pieces in four different sizes. This action is meant for the protection from the sun depending on its direction and for the different looks to the pottery factory.

项目名称：Triana Ceramic Museum
地点：31 San Jorge Street, Sevilla, Spain
建筑师：Miguel Hernández Valencia, Esther López Martín, Juliane Potter, Francisco José Domínguez Saborido, Ángel González Aguilar
合作商：Angélica Cortés, Ana Blanco(competition), Elías Pérez, Rubén Ingelmo, Reyes López(project)
测量师：Reyes López, Rafael Esteve, Rafael M Esteve
考古学家：Miguel Ángel García
博物馆学家：Alfonso Pleguezuelo
修复师：Alfeizar
施工单位：Ute Condisa-Alea Global, Espai Visual (Museology)
甲方：Consorcio Turismo de Sevilla
用地面积：1,501m²
有效楼层面积：1,052m²
总建筑面积：2,227m²
项目时间：2009—2010
施工时间：2011—2012
摄影师：©Jesús Granada

面向Campos大街的立面 elevation from Campos Street

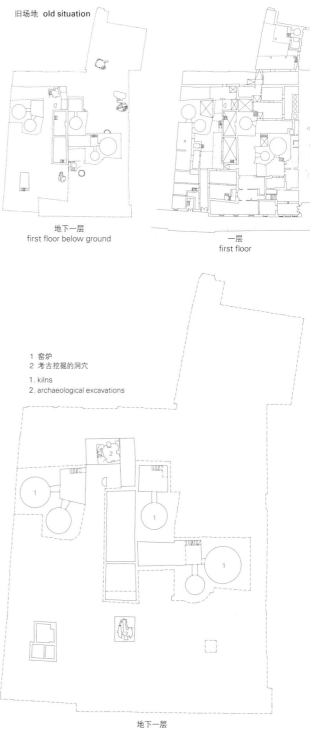

旧场地 old situation

地下一层 first floor below ground

一层 first floor

1 窑炉
2 考古挖掘的洞穴
1. kilns
2. archaeological excavations

地下一层 first floor below ground

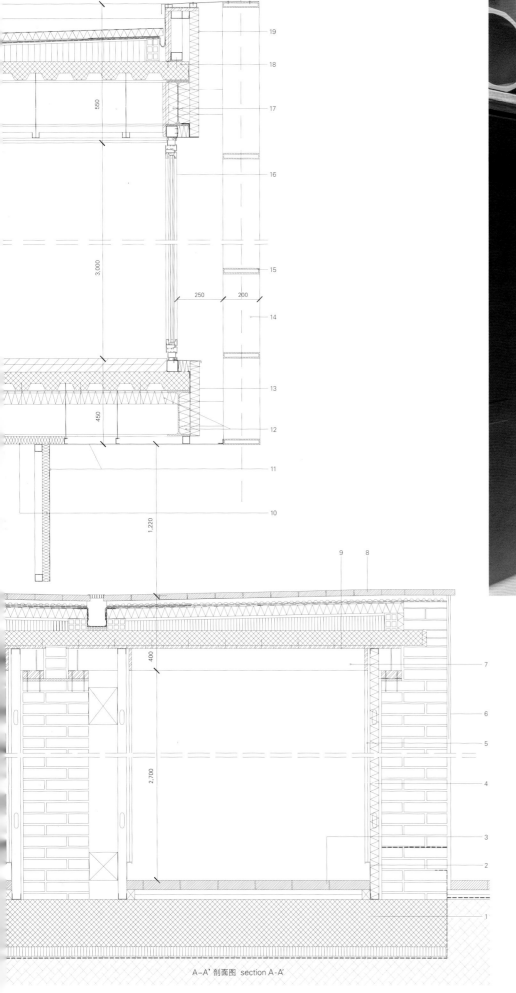

A-A' 剖面图 section A-A'

1. foundation
2. steel sheet skirting (2.5mm)
3. ceramic paving stone
4. thermal insulation and sound proofing (mineral wool 40mm)
5. free-standing laminated plaster boards
6. lime mortar (20mm) + silicate paint
7. laminated timber beams
8. floor of ceramic bricks
9. wooden board
10. sandwich panel for sound proofing with perforated metal on the exterior side and sound proofing on the interior side (on the technical floor).
11. perforated metal (2mm)
12. extruded polystyrene insulation board (50mm)
13. sandwich panel (50mm)
14. vertical steel post #200x30x3mm
15. horizontal steel shelves #200x30x3mm
16. aluminum window
17. panels for thermal insulation and sound proofing (mineral wool 2x50mm)
18. structural steel
19. sandwich panel (50mm)

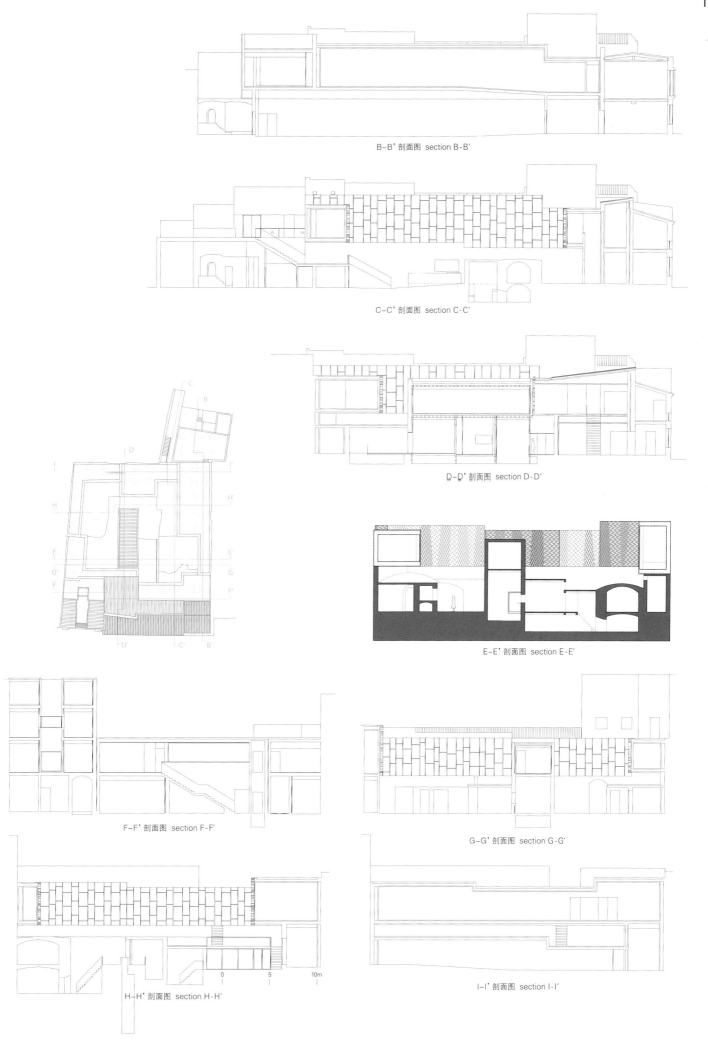

YOONGYOO JANG + UnSangDong建筑师事务所

YoonGyoo Jang + UnSangDong Architects

社会想象

我们生活在这样的一个时代，即建筑想象力是非常荣耀的。在形式建造和空间建造中，我们沉迷于更新的创意和独特性。想象力加速地产生，但是却由于个人主义的出现而导致其受到一定的限制。今后，我打算对潜在的想象力加以利用，将其作为改造城市及社会的工具，旨在期待由社会想象力建成的新城市和新建筑。

基本上，我认为想象力将会创建一种全新的定义建筑并且将建筑与其相关区域相融合的方法。整体的工作始于质疑，质疑传统的空间理念，或者是质疑定义空间的具体构件的理念。这是一种全新的、对过去未确认的建筑进行探索的尝试。

我对建筑的空间和形式做了若干次实验，这些实验包括建造区域、改变功能区以及探索建筑本身的意义。通过实验，一些建筑中持久存在的价值得以重塑，并且体现在各式各样的位移和变形中。UnSangDong建筑师事务所便是在想象力的基础上成立的，这个事务所利用创意、无边界的城市、百科全书、复合体、反应物等进行了实验，做出了很多的努力，并且已经走在了建筑的前沿。这些实验在城市环境和严谨的环境可持续性检验中重建了序列，其结果不仅仅是一个终端性成果，而且还建立于我努力求得的新型乌托邦想象中。

我感兴趣的是代表自然集体记忆的东方景观，同西方景观相比，西景观旨在展示眼睛所及的现实。东方景观通过覆上一层艺术家的想象力（对于大自然来说不是正确的视觉观点）创造了梦幻般的场景。Ahn Gyeon是朝鲜王朝时代的画家，重建了Mongyudowondo画作中AnPyong王子的梦。这处带有异域风情的乌托邦式景观位于一座道路崎岖、充满岩石的山上，其景色部分是现实，部分是想象。这位画家通过他对Mongyudowondo的想象来进行创作，而在这幅画中，这座岩石山的真实景观与脑海中的乌托邦式想象融合在一起。也许，Mongyudowondo就是想象到达一个极限的边缘，且是超越了我们所熟悉的一般想象的界限的典型例子。

Social Imagination

We live in an age where architectural imagination is glorified. In form making and space making, we are addicted to even newer originality and exceptionality. As the imagination grows in acceleration, however, there comes a limit due to individualism in a society. Henceforth, I intend to use the potential imagination as a tool to transform cities and societies. I aim for new cities and new architecture created by social imagination.

Basically, I believe imagination will create a new way of defining and integrating architecture and its relevant areas. Overall my work begins in questioning, questioning traditional spatial concept or questioning conceptual aspects to physical elements which define space. It is an attempt to newly discover architecture un-identified in the past.

I have made experiments in space and form making areas, deforming programmatic areas, and modifying architectural significance. Things considered as constant value in architecture have been re-created into various displacesants and distortions through the experiments. UnSangDong Architects has been established armed with imagination, experimenting with originality, clip city, encyclopedia, compound body, reaction body and etc., and has made many efforts being at the forefront of architecture. The experiments reconstruct order in urban environment and seriously examine environmental sustainability, and the outcome is not only an end result but further builds on the new utopia of imagination which I work to achieve.

I am intrigued by the Eastern landscape which "represents collected memories" of nature compared to the Western landscape which aims to "exhibit the reality" seen by the eyes. The Eastern landscape creates dreamlike quality as a result of adding a layer of the artist's imagination which can be an inaccurate view of the subject nature. Ahn Gyeon, the painter of JoSeon Dynasty, has re-created Prince AnPyong's dream in "Mongyudowondo". The landscape of an exotic utopia realized a rugged rock mountain panorama made of reality and imagination. The artist has created his imagination "Mongyudowondo", where two unsuit-able set up, the realistic landscape of rock mountain comes together with

Water Circle博物馆_CheongShim水文化中心_Water Circle – CheongShim Water Culture Center
文化街——SeongDong文化和福利中心_Culture Street – SeongDong Cultural & Welfare Center
运动想象——2012韩国YeoSu世博会现代汽车展馆_Motion Imagination 2012 – YeoSu EXPO Hyundai Motor Group Pavilion
Rooftecture——能量+绿色住宅_Rooftecture - Energy + Green Home
由复合体向社会想象体的转变_Transformation of Compound Body to Social Imaginative Body / YoungBum Reigh

我所提出的想象便是如同Mongyudowondo中的一样,将"虚构的想象力"与"社会想象力"结合起来。"虚构的想象力"会充分挖掘潜在的想象力,这是以科学为基础的想象力无法企及的。创造与克隆,组合与结构,转变与改造……我认为城市的视野充满了各种可能性,没有界限。起初,通过观察社会和城市现象,我的工作风格就是对相互关联的元素进行提取和组装,如探寻超越了对立性限制的魔术。不仅仅如此,对于科技带来的所有可能的界限模糊性,我还致力于探索其内在的潜力。但是这种想象力存在的前提是其必须存在于一个社会内,因为城市和建筑都是在其内产生的,并且反映了这个时代。

这种由城市增加的人口、建筑和设施所引起的扩张正在对城市的领域和食品提供造成威胁。这个成本驱动的社会已经对环境的消耗到达了极限,像全球变暖、生态破坏等等。这些危机要求"社会想象力"能够在城市和建筑区域产生可持续性。而这种想象力的界限可能始于无法认识到社会的需求。社会想象力成为城市和环境之间进行互动的一种工具,并且有可能产生重要意义。这一工具在一种全新的方法中负责建造城市设施。"社会想象力"通过各式各样结合城市元素的方法,来质疑最后的结果。不仅如此,它还为不同城市元素结合所产生的建筑空间提供一个基础框架,并且还影响着这处空间。"社会想象力"实际上已经超出了界限,并且将城市和建筑领域强有力地融合在一起。这是我们这个时代亟需的一种期望,以激发新的想象力,去创造新型城市结构和空间。

a utopia of minds imagination. Perhaps, "Mongyu-dowondo" is an example of imagination reaching its extreme edge which goes beyond the limits of general imagination we are familiar with. The imagination I propose is such as in "Mongyudowondo" which joins together "mythological imagination" and "social imagination". "Mythological imagination" operates to fully reach the potential of imagination whereas science-based imagination of our times has failed. Creating and cloning, uniting and deconstructing, transition and transformation... my vision of cities are full of possibilities without limits. Initially making observation of social and urban phenomenon, my methodology of working style is extracting and uniting inter-related elements, such as seeking magic that exceeds the limits of contradiction. Not only that, I seek to explore potentials of fusing all possible limitations caused by technology. The premise of this imagination only exists within a society. Because cities and architectures are born in and reflect the periods.

The expansions caused by increasing urban population, architecture and infrastructure are threatening urban domestic territories and food supplies, and furthermore ecological environments. The cost-driven society has reached the limits in environmental depletion such as global warming, ecological damage and etc.. These crises demand "Social Imagination" to reach sustainability in urban and architectural field. The limits of imagination may have started from not being able to realize the needs of the society. "Social Imagination" becomes a device to bring interaction between cities and environments, thus creating possibilities in new significance. This device performs a role of creating urban infrastructure in a new method. "Social Imagination" continues to question the outcome by means of various methods of combining urban elements. Not only that, it provides a fundamental framework of role and influence of architectural spaces created by combining different urban elements. "Social Imagination" truly goes beyond the limits to powerfully combine urban and architectural territories into one outcome. This is an expectation necessitated in our times to inspire new imagination to create a new urban structure and space. YoonGyoo Jang

由复合体向社会想象体的转变
更具有社会性,并且向往建筑具有更多的道德性

Transformation of Compound Body to Social Imaginative Body
More Social, therefore Dreaming for More Ethics of Architecture

向建筑具有更多的道德性方向发展

如果建筑有可能意味着一种出售丰富人类生活价值的可能性,那么我们要询问的最重要问题之一便回归到建筑本身,即"你怎么使空间获得应有的道德?"为了获得最小程度的道德伦理,建筑被视为文化霸权,并且注重建造过程与态度。在2000年,即21世纪的初始点,威尼斯双年展要求在全球范围内与建筑相关的有趣的主题都要进行一个自省。那么这一主题"城市:少一点美学,多一点道德"蕴含的意义是什么?它对原创者的道德放纵提出了严肃的警告,这些原创者很轻易地便妥协于资本和自大的精英主义(其拥有强加给社会的、以主流价值观为基础的文化代码)。然而,因为建筑被局限在相互冲突的构成元素——资本与社会、概念与施工、空间与事件中,因此它便要在道德和美学的范畴内,承受走在钢索上的固有限制。

在城市的背景下,由资本流动所引起的连续的循环占领了主导的地位,成为一种绝对的价值,意义大过其存在的本身。利用城市(任何事件都处于不稳定的状态)中的建筑语言来追寻空间价值的根源,这种做法看起来是十分荒谬的。尽管如此,在改造的城市和不断变化的社会中存在的建筑价值则取决于一种空间态度,这种态度倾向于如何改变其背后的力量,且与这种力量相互影响。基于作者的社会性,建筑能否将美学与道德结合起来?在本文中,我将会看看UnSangDong建筑师事务所的项目是怎样考虑这一问题的。

复合体

复合体是过去的十年间UnSangDong建筑师事务所项目的基础理念。复合体,将建筑理念付诸于实践,并且做空间实验,可以被认为是接近作者视角的一个关键词。然而,在实现梦幻般的、想象中的世界时,这一处理还是有潜在的危险的。作为一种释放意愿的方式,建立复合体的理念是非常有用的。UnSangDong建筑师事务所的项目探索建筑与城市之间的界限、现状与虚拟的场景、空间与人类表现,其重要性存在于通过复合体建筑来体验各种边界交叉的规划空间中。当UnSangDong建筑师事务所拒绝单一的规划文本,并且通过改革与操作来创造分层的空间规划时,他们集中采用可操控的超文本式空间来取代单一文本式的建筑表皮和实体。复合体、空间想象体和认知超越了其物理存在,在这一超文本中可以被看做是一个连续体,它将认知空间的限制从体验中分离开来。如果是这样,为什么UnSangDong建筑师事务所沉浸于如此的工作中?当复合体的理念被理解为一种曝光的、社交的且被社会扭曲的空间代码时,它便具有了有效性。只有当空间被设计成一种有过经历,而非

Towards More Ethical Architecture

If architecture connotes the possibility as the act of selling the values which enrich human life, one of the most important question we ask return to architecture would be "how do you obtain ethics of space?". In order to obtain minimum ethicality, architecture regarded as cultural hegemony, pays attention to architectural process and attitude which social discourse actualize into space. In 2000, at the starting line of the 21st century, Venezia Biennale demanded global introspection of architecturalization with interesting subject. What signification does connote in the subject "The City; Less Aesthetics, More Ethics?" It has a serious warning against Moral Laxity in authorship which easily compromises with capital and Arrogant Elitism which own cultural code based on subjective values imposed on the society. However, as architecture is confined in the conflicting composition of capital and society, concept and construction, space and happening, it bears inherent limitation walking on a tightrope at the boundary of ethics and aesthetics.

In the urban context, continuous cycling of the existence caused by capital flows dominates as an absolute value rather than the existence itself. It seems absurd to ask the root of spatial values through architectural language in the urban where nothing is stable. Nonetheless, architectural values within transforming cities and endlessly changing society depend on spatial attitude towards how to react to the changing and interact with the force behind it. Is it possible to combine aesthetics and ethics through architecturalization based on authorial perspective on sociality? Here I would like to look around UnSangDong Architects' projects considering of this question.

Compound Body

Compound Body is a concept which forms the basis of UnSang-Dong Architects' projects for the past 10 years. Compound Body, practicing architectural concepts and spatial experiments, can be seen as a keyword close to authorial perspective. There is a probable danger in dealing with realizing dreaming imaginary world. As a way of erupting their desire, establishing the concept of Compound Body has been much effective. The significance of UnSangDong Architects projects, which exploring into the boundaries between architecture and urban, actuality and virtuality, and space and human performance, lies in proposing space that is allowed to experience cross between various boundaries through architecturalization of Compound Body. As UnSangDong Architects refuses single text and thus creates spatial layering of programs by transformation and manipulation, they are focused on substituting manipulated hypertext-like space for skins and objects as individual texts. Compound Body, spatial body of imagination and recognition surpassing physical existence, can be interpreted as continuity in hypertext which breaks the limits of recognized space from experience. If so, why has UnSangDong Architects been immersed in such work? The concept of the Compound Body has validity when understood as the process of exposing and socializing spatial code distorted by the society. Only if the space is designed as an object to be experienced instead of objectifying it, and occupants are considered the main agent to not observe but experience, it is possible to become architecture as a compound of human and space. Also, reproduction of unreality, which provides experience of unreal world by space,

客观化的物体时，其居住者才会被认为是其主要的代理人，而代理人要去经历这处空间，而非观察它，那么将人类与空间进行复合也就变得可能了。此外，不真实性的复制通过真实空间来为非现实世界提供经历，且通过复合体来实现。这里，社会属性和想象成为这一过程中的两个重要的主题。当建筑师的非现实想象以上述的两个主题为基础，在现实生活中被富有创意地解读时，那么不稳定性便由此产生了，这将会激发UnSangDong建筑师事务所的创造力，使其作品集中在复合体上面，以作为一个超文本，连接关系网的接触点。

社会想象

城市中关系网以及无数的空间关系的多样性，如适应、冲突、共存、自我繁殖等，本身就是一个概念团体。但当我们仔细地探寻其中时，这种在城市中所展现的多样性便仅仅是同质性和异质性组成的二分法所形成的改革和操作。这就是为什么我们会对城市边界问题感到矛盾和令人窒息的原因，城市边界是我们利用与空间的关系来进行交汇的地方。最后，这种相互冲突的矛盾使一处新型空间规划成为可能。空间方面的新想象力，取决于一些可能性是怎样从各式各样的未指明的空间、建筑以及城市中分离开来的。然而，空间想象并不仅仅依赖于建筑师的概念想象。建筑通过无数的潜藏在空间的物理、技术和社会元素，如文化、科学、技术、运输以及其他更多的，来使其本身成为城市与社会的交界面。向前走一步，它可以通过需求，利用技术，以连续的历史和文化为基础，来建造一处富有变化力的空间，它与复合体是有直接的联系的。那么，我们会再一次地发问："复合体的基本设计目标是什么？"如果复合体是一个超文本式的空间，那么它会面向所有未定的和未完成的建筑可能性开放，而超文本的范围使这种模棱两可性作为一种加速质疑空间本质的战略，达到了最大化，同时也使空间与社会之间的交界面（非确定）所产生的可能性也提高到一个顶点。我们生活的城市（城市化的社会）是一些不确定性共存的地方，甚至对我们所信仰的，将其作为本质的信念产生威胁。UnSangDong建筑师事务所没有对具体的事物进行观察，以对城市中的建筑进行学习，而是注重城市中潜在的无形的联系。看待具体事物之间存在的物体的态度是一种存在于社会想象体中的可能性。事实上，能够完美地展现理念和物质的本质，是一件不可能完成的任务。但是，想象是其有可能实现的力量。不以材料属性为依据来展示其实质，也不以作者的视觉来重塑理念，而是将关系网的想象空间化，使其成为可以被无限阅读的文本，以提供更多实质性的东西。

本文中所介绍的四个项目都是从重塑和建筑的角度出发，以成为

is achieved through Compound Body. Here, social nature and imagination become two important topics in this process. There is instability here when architect's unrealistic imagination can be reinterpreted creatively in reality based on two topics mentioned above. And this motivates creative force in UnSangDong Architects' works, which concentrates on Compound Body as hypertext and contact points along with relationship network.

Social Imagination

The diversity of relational network, numerous spatial relations in city such as adaptation, confliction, coexistence, self-generation, and etc., is itself a conceptual lump. But as we carefully look within, the diversity expressed in cities are only transformation and manipulation made by dichotomy of homogeneity and heterogeneity. Even though it is the reason why we feel confliction and stifle at the urban boundary where we meet through the relationship with space, eventually this conflictive composition makes possible to propose a new type of space. New imagination regards to space, depending on how possibilities driven from various boundaries of unspecified spaces, architectures and cities, present itself will determine the outcome. However, spatial imagination doesn't merely rely on architect's conceptual imagination. Architecture reveals itself as an interface between cities and societies through countless physical, technical and social elements implied in space such as culture, science, technology, transportation, and many more. To take a step forward, it can create transmutative space by needs with technology based on continuous history and culture. It is in association with the Compound Body. "What is the fundamental aim of the Compound Body?" We shall ask again. If Compound Body is a hypertext-like space, it is opened to make undetermined and uncompleted potential in architeture, then makes the boundary of the hypertext maximize ambiguousness as a strategy to accelerate questioning the essence of space and also maximize possibilities made from uncertain interface between space and society. The city we live in (urbanized society) is where uncertainty co-exists, even threatening our belief of what we believe as the essence. Instead of observing physical objects in order to study architecture of the city, UnSangDong Architects pays attention to invisible relations which is immanent in the city. The attitudes towards something existed between the physical subjects are possibilities beared in Social Imaginative Body. In fact, it is an impossible task to perfectly reveal the essence in concept and substance. But Imagination is energy to make it feasible. Instead of revealing the essence based on material property or regenerating concept with authorial perspective, but spatializing the imagination of network read as limitless text will provide more than the essence.

4 projects introduced here are set off from regeneration and architecture as an environmental machine. When creating circulation system of relations within space, architect's concept is regenerated by human activity in space. The regeneration mentioned here, identifies to conceptualized spatial essence being regenerated in the form of relational network through human activity based on instant demand. When regenerating values, it is important to propose a circulation system which induces change in new urban environment, rather than aesthetical manipulation reached

一架环境机器。当建筑师在空间内创造一个循环的关系系统时,他们的理念也被空间内的人类活动进行重塑。这里提到的重塑,即重新提出的概念化的空间本质,以人们的即刻需求为基础,通过人类的活动展现出来,并且其形式体现为相互联系的关系网。当价值得以再生时,那么规划一个循环系统是十分重要的,相对于改变审美操纵(由建筑目标来达成),这个系统更易于在新城市环境中发生变化。这一循环系统允许理念和意义、空间和活动、关系和性能不断地进行重塑。当系统蕴含在空间时,建筑则成为一个反应物。因其是一个反应物,而非一件已完成的产品,那么响应对象之间的关系网便由此产生。这个关系网使建筑与社会之间的界限开始模糊。这就是建筑是一个有机体,而非终结的社会反应的原因。

文化街——SeongDong文化和福利中心

在SeongDong文化和福利中心项目中,最重要的问题不是将其建成一座纯粹的政府办公式建筑,而是使这座文化和福利中心在贫瘠的城市环境中能够行使管理的功能。项目位于环境恶化的工业区中心,扮演着改造城市的首脑角色。这个为公众服务的办公室主要行使行政的功能,从另一方面来说,管理机构的建筑空间可以被视为关键角色的种子,意图改造城市与社会。SeongDong文化和福利中心便是发挥着协助与重新规划工业区的居民生活的功能。外部街道通至该区域,以作为垂直交通路径,建筑轮廓与表皮成为自然与社会之间的接触点。SeongDong文化和福利中心连同格拉茨的美术馆和伦敦的Peckham图书馆一起,从建筑能够改变城市与社会的前提出发,将重塑社会交通系统的功能与连接行政管理职能和文化福利的功能融合在一起。此外,项目还为规划公共空间提供了可能性,如广场和城市绿化设施,这一事实介于城市和建筑领域的边缘,在社会想象体的背景下能够得到充分的理解。

Water Circle博物馆——CheongShim水文化中心

这个项目是一个净化不同种类的污水的基础设施,这些污水都是来自于CheongShim开发综合体建筑。有趣的一点是,这个项目除了是一座净化污水的设施外,还是一种净化心灵的建筑体验。CheongShim Water Circle博物馆让人们理解水的净化过程,这种体验类似于艺术博物馆,结合了被忽视的污水设施,并且展示了其成为社会设备,且作为"环境基础设施和水讲解工具"的改造过程,人们在这里可以经历到各式各样的体验。作为一种生命循环的建筑象征,这座建筑通过人类和自然之间的互动,重新定义了水、自然和人类的关系。此外,它还通过刺激五官,来设置用于看、听、享受和交际的体验空间,这些都是由社会想象体旨在创造的价值。

by architectural intention. This circulation system allows endless regenerations of concept and significance, space and activity, relation and performance. When this circulation system is implied in space, architecture becomes a reactant. Being a reactant instead of a finished product, many networks between responsive objects are created. The networks have made the boundary between architecture and society ambiguous. This is the reason why architecture acts as an organic body not a terminated reaction of the society.

Culture Street – SeongDong Cultural & Welfare Center

In SeongDong Cultural & Welfare Center, the important issue is not making a mere government office type building but a governance role performed by cultural&welfare center appearing in poor urban environment. Located at the very center of deteriorated industrial area, the cultural & welfare center plays a role as headquarter transforming the urban. The function of public office is focused on administrative role, in other hands, architectural space created by governance can be regarded as seeds of pivotal role intended to transform cities and societies. SeongDong Cultural & Welfare Center performs a role of assisting and reorganizing the life of residents in the industrial area. External street was brought in as the vertical path and composed its shape and ecological building skin is a contact point between nature and society. SeongDong Cultural & Welfare Center, compositing the role of regenerating social circulation system and spatially articulating the governance of administration and cultural welfare, starts from the premise that architecture can transform both city and society, along with Kunsthaus in Graz and Peckham Library in London. Also, being at the border of city and architecture, the fact that it proposes the possibilities of public space such as plaza and urban green infrastructure is comprehended in the context of the Social Imaginative Body.

Water Circle – CheongShim Water Culture Center

This project is an infrastructure which purifies different sewages made from all over CheongShim development complex. The interesting point of this project, above functional facility for sewage purification, is architectural experience transformed into spiritual purification. CheongShim Water Circle allows the process of purification of water to be understood similar to art museum experience which combines sewage facility looked down upon, and also provides transformation into social device as "environmental infrastructure and water education" where various experience of water can be made. As an architectural symbol for the circulation of life, it redefines the relationship of water, nature and human by interaction between human and nature. Additionally, it generates experiential space of looking at, listening to, enjoying, socializing with by stimulating five senses. And these are the values aimed by Social Imaginative Body.

Rooftecture – Energy + Green Home

As one theme of new housing types, this project performs Energy + Green Home which exceeds Energy Zero. This project is a

Rooftecture——能源+绿色住宅

作为新住房类型的一个主题,这个项目成为优于零能耗的能源+绿色住宅。这一项目通过结合景观建筑和生态建筑,成为一个自然与人类生活共存的可持续性住宅。项目的主要建筑意义在于覆盖了整体体量的一个连续的屋顶结构。建筑中再生循环系统在这个住宅类型中得以实现。这样的建筑对于社会来说,仍然是一个非响应的物体,旨在其本身的竣工。但是它却利用自然,将循环系统包含其中,来作为内部的空间序列,形成一个有机体(包含无数个对环境做出反应的交界面)。

交流体与价值得以重塑的环境&社会想象体

在复合体延伸进社会想象体的同时,文本共享、价值重塑、主要和创新领域的整合是UnSangDong建筑师事务所对于建筑的一贯态度。上述的项目作为建筑时并不能运行,而是要与周围的社会需求发生反应。它们都是较为主动地,能够接收无限的信息,在关系网中利用社会进行改造,而非被动地,仅仅依靠建筑师的理念来运行。由建筑所创造的文明象征则代表着环境视角内的废墟。所有的建造的空间系统都使自然环境成为废墟,这些都是由更加文明的社会造成的。在这一点上,"本质是什么",这一问题就变得十分必要。在UnSangDong建筑师事务所内交流的这一主题便超越了复合体的范围,进而进入了社会想象体的范畴,以重塑价值。这一项目便是探求与建筑相关的社会需求的实质。首先,项目中所需的态度便是关注存在于主体(体验建筑,将建筑与之结合)和客体(是建筑无休止地利用空间来产生的)之间的反应。你通过什么样的方法来展示关系网中无法计数的隐形关系?你怎样在空间里展现最接近本质的环境?根据这些问题的答案,建筑被改造成由大量的关系网编织而成的环境。从人类的整体机构、建筑、城市及构成它们的构件来说,我们生活在由室内和室外关系来定义的动态结构中,最终赋予建筑的任务仅仅是建造一个连接这些关系的设备。UnSangDong建筑师事务所的建筑视觉相信以新颖性为基础,为建筑重新定义并且将建筑与其相关的领地相融合,是有可能的。他们的工作始于质疑,可能从传统的空间见解开始,横跨概念部分和具体构件,他们想探索过去的各种未经鉴定的建筑。UnSangDong建筑师事务所通过他们连续建成的项目(包括上述的四个项目)来展现其一贯的建筑态度,并且展示了能够重塑价值的社会想象体,而重塑的价值是由作者天马行空的现象力和社会想象力结合技术来实现的。

sustainable housing for co-existence of nature and human life by combining landscape architecture and ecological architecture. The fundamental intention of architecturalization in this project lies in the continuous roof structure consisting of the overall mass. Circulation system of regeneration in architecture is actualized here in housing type. This architecture with the concepts stands still as an non-reactant to society fulfilling self-completion, but it embraces the circulation system with nature as internalized spatial order and completes the organic body that consists of countless interfaces reacting to the environment.

Communicating Body with Environment & Social imaginary Body of Value Regeneration

Text sharing, regeneration of value, integration of domain and innovative space are UnSangDong Architects' consistent attitude towards architecture whilst Compound Body extends to Social Imaginative Body. These projects above not only survive itself as architecture, but also react to social demands surrounding them. They are active subjects receiving infinite information and transforming in network with society rather than passive objects dependent on the architect's concept. The evidence of civilization made through architecture signifies a ruin in environmental perspective. All constructed spatial system makes the natural environment a ruin due to ever more civilized society. At this point the question of "what is the essentiality?" is necessary, the topic of conversation at UnSangDong Architects goes beyond the Compound Body and into Social Imaginative Body in order to regenerate values. This work is asking the substance of architecturalization society demands. Firstly required attitude in this work is paying attention to reactions existing between the subjects who experience and combine architecture and the objects architecture endlessly generates through space. How do you reveal endless invisible network relationship? How do you spatially perform the closest circumstance to the essence? Depending on the answers to these questions, architecture transforms into circumstances made of numerous network relationship. From the whole set-up of human, architecture, city and to its elements composing it, our life exists within a dynamic structure defined by internal/external relations. Eventually the task given to architecture is merely creating a device interconnecting these relations. UnSangDong Architects' architectural perspective believes that basically through newness, it is possible to make new definition and integration of architecture and its related territory. Overall their work begins in questioning, from perhaps traditional spatial notion to across the conceptual part and physical element, they want to discover various architecture un-identified in the past. UnSangDong Architects has shown consistent attitude of architecturalization through their continuous work throughout including 4 projects introduced above. It reveals Social Imaginative Body which regenerates values realized by authorial mythological imagination and social imagination combined with technology.

YoungBum Reigh

Water Circle博物馆
CheongShim Water Culture Center

YoonGyoo Jang + UnSangDong Architects
YoonGyoo Jang + UnSangDong建筑师事务所

人类生活就是一个每天消耗大量的水的累积过程。水是一种宝贵的资源，与生命息息相关，没有任何其他资源可替换。然而，在人们创建文明的城市和乡镇的同时，也造成了污染，浪费了大量的水。因此，一项重要的任务已经摆在了人们面前，有必要建立污水净化系统，努力减少污染，循环利用水资源。

一般来说，建立污水净化设施时会特别关注污水净化这一功能，不会有任何环境艺术和美观方面的考虑，其外观通常不会给人令人愉快的美感。然而，Water Circle将这一往往不美观的设施与艺术体验结合起来，使人们了解真实的净水工艺，分享各种利用水的体验。所以，Water Circle成为一个社会工具，让人们对环境基础设施和水有更好的了解。换句话说，Water Circle实际上用来进行污水净化，同时作为建筑，还承担着净化社会和心灵的作用，让人们意识到水的重要性。这一富有创造性的基础设施通过相关而复杂的功能，如教育、展览、环保设备和自然空间等，来展现自然的活力。

水在大地间流淌

水是孕育整个世界的最重要元素，通过水元素，本设计努力展现坚忍不拔的毅力、绝对性和直观的建筑。

设计采用最原始的几何形状——圆形，使建筑能够体现生命的纯洁和最初如水晶般的清透。大地被切割成几块，形成完美的几何图形，即圆形。圆墙直径为32m，高度是11m，为了展现隐藏于地下的水的活力，建筑被提升，高于地面，希望通过高高悬浮于地面的曲形墙来表示地球与建筑之间最简单、最静默的规则。建筑内部采用了台阶这一抽象的建筑语言，向参观者展示了体验水以及了解污水净化过程的地下空间。曲形阶梯式地表之间展现了水的各种形式。建筑师希望这一建筑像水一样流动起来，在沉默宁静中展现水的生命力。

污水净化和展览之间不断变化的空间

外立面墙呈绝对的圆形，墙体上建的台阶也是建筑语言，表达出水在内部和外部空间流动的穿透力。外部台阶状的空间是围墙内的中空空

间，也是人们心灵上与大自然、周围景色和风亲密接触的空间。

室内展览由两部分组成。一是地下空间，实际上是进行污水净化的地方。在这里，参观者可以看到真正的通过机器进行污水净化的过程。这一展览可以让人们了解水的重要性，重建新的环保意识。这是现场污水净化展览不可或缺的空间，承担着教育、展览、体验三种角色。

水往低处流，上层较高的展览空间与一层和二层是连为一体的，展现出水各种各样的形式，让参观者对水有一种全新的体验。在设计中，建筑师强烈要求在建筑内部创建水环境，让参观者来体验水。在贯穿一层和二层的中空空间中，人们可以体验各种各样富于变化的水像雨一样从天而降，也可以通过在落入一楼池塘的水里面嬉戏来体验大自然。当然，参观者也可以通过互动媒体与水沟通交流。在烟雾笼罩的玻璃箱中，参观者可以体验雾中漫步。为了让参观者五官都能对水有不同的感受，所有展览区都提供了一系列不同的体验项目。

二楼增加了教育、展览、图书馆和供参观者休憩的空间。屋顶成为参观者欣赏周围一览无余的优美景色的地方。

Water Circle

Human life is a cumulative set of daily water consumption. Water is a valuable resource and there is no other replaceable resource associated with life. However, people have caused the pollution and wasted the water while they created civilized towns and cities. Therefore, an important task has occurred. It is necessary to build purification systems and make efforts on reducing the pollution and reusing the water.

Generally, sewage purification facilities are built focused on the function as purifying polluted water, they are generally created as unpleasant facilities without any environmental and aesthetic considerations. However, Water Circle has combined the unpleasant facility with art experiences to understand the actual water purification process and share various experience with water. so it transforms to the education of environmental infrastructure and water as a social device. In other words, Water Circle actually oper-

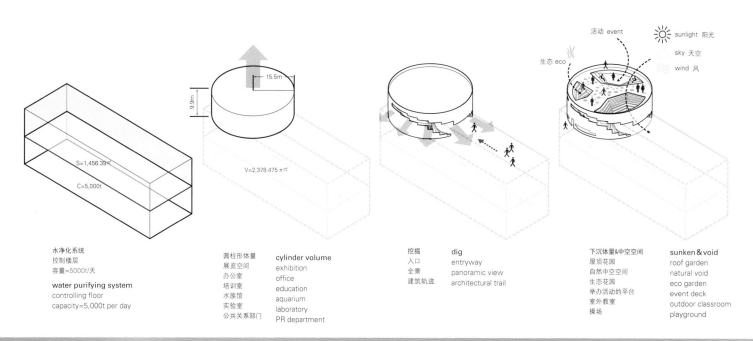

水净化系统	圆柱形体量	cylinder volume	挖掘	dig	下沉体量&中空空间	sunken & void
控制楼层	展览空间	exhibition	入口	entryway	屋顶花园	roof garden
容量=5000t/天	办公室	office	全景	panoramic view	自然中空空间	natural void
water purifying system	培训室	education	建筑轨迹	architectural trail	生态花园	eco garden
controlling floor	水族馆	aquarium			举办活动的平台	event deck
capacity=5,000t per day	实验室	laboratory			室外教室	outdoor classroom
	公共关系部门	PR department			操场	playground

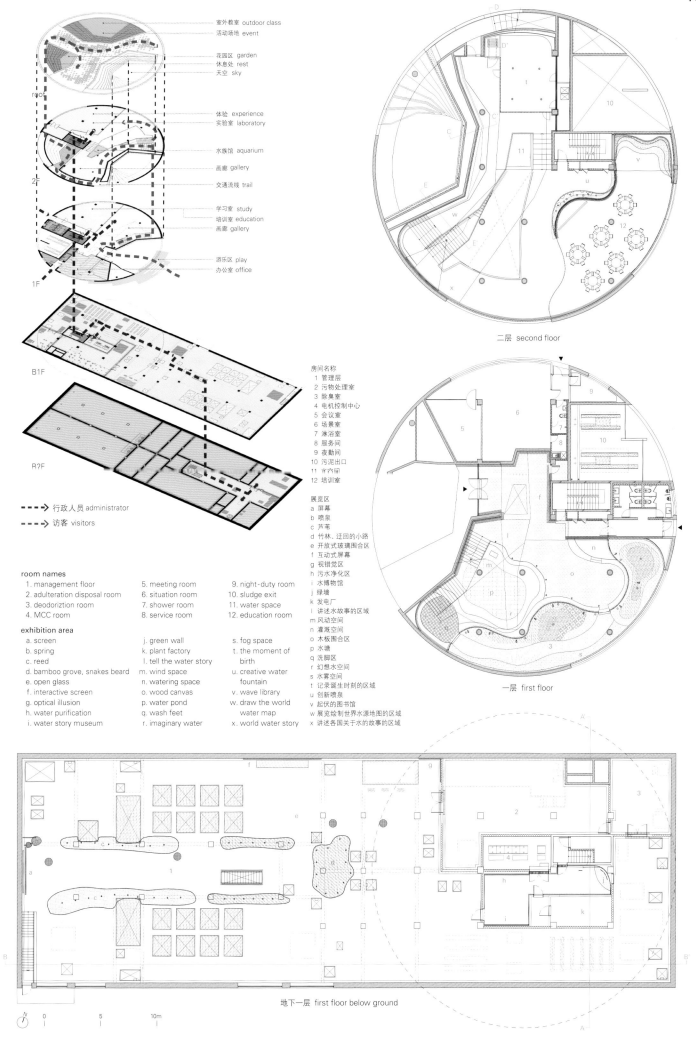

room names
1. management floor
2. adulteration disposal room
3. deoderiztion room
4. MCC room
5. meeting room
6. situation room
7. shower room
8. service room
9. night-duty room
10. sludge exit
11. water space
12. education room

exhibition area
a. screen
b. spring
c. reed
d. bamboo grove, snakes beard
e. open glass
f. interactive screen
g. optical illusion
h. water purification
i. water story museum
j. green wall
k. plant factory
l. tell the water story
m. wind space
n. watering space
o. wood canvas
p. water pond
q. wash feet
r. imaginary water
s. fog space
t. the moment of birth
u. creative water fountain
v. wave library
w. draw the world water map
x. world water story

项目名称：CheongShim Water Culture Center
地点：SongSan-ri SorAk-myeon GaPyeong-gun GyengGi-do, Korea
建筑师：YoonGyoo Jang, ChangHoon Shin, KyoungTae Kim
项目团队：SamYeol Ryu, BongKyun Kim, SeungHyun Kang, JeHyun Sim, MinSun Son, ByungWoo Kim
结构工程师：Thekujo 机械与电气工程师：HIMEC
室内设计：USD
甲方：CheongShim
用地面积：3,220m² 总建筑面积：671.65m² 有效楼层面积：2,608.61m²
结构：reinforced concrete
室内设计时间：2010.12—2011.6 室内施工时间：2011.6—2012.1
施工时间：2012.1—6 施工时间：2012.7—2012.12
摄影师：
©JongOh Kim - p.134~135, p.136~137, p.138, p.140, p.141, p.142~143, p.144[bottom], p.145[top], p.146[bottom-left], p.147, p.148, p.149
©Sergio Pirrone - p.144[top], p.145[bottom], p.146[top], p.146[bottom-right], p.150~151

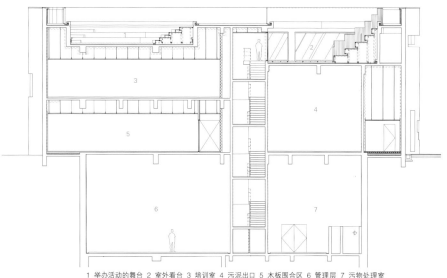

1 举办活动的舞台 2 室外看台 3 培训室 4 污泥出口 5 木板围合区 6 管理层 7 污物处理室
1. event stage 2. outdoor stands 3. education room 4. sludge exit 5. wood canvas 6. management floor 7. adulteration disposal room
A-A' 剖面图 section A-A'

ates and purifies sewage, it also educates the importance of water as an architecture which forms social and spiritual purification. This creative infrastructure contains the vitality of nature through associated complex functions such as educations, exhibitions, environmental equipments, and natural spaces.

Water Flow between Earths

Through the "Water", which is the most fundamental element which has conceived the world, unwavering tenacity, absoluteness, and intuitive architecture have tried to be discovered.

The most primal geometry, circle, generates architectural realization as a pure and primal crystal of life. The land is sliced into geometrically perfect shape of circle with the wall of 32m diameter and 11m heights, and it is lifted in order to discover vital power of water hidden under the earth. The relationship between the earth and architecture wants to create the simplest and silentest order through the curved wall soaring high above the ground. Inner land is abstracted to step-shaped architectural language and provides underground space to experience water and its purification. Various forms of water have been contained between the curved stepped earths. The architects suggest the architecture of flow, which is associated with the vital power of water in the silence serenity.

Flux Space between Purification and Exhibition

The stepped ground in the wall of absolute circle is also the architectural vocabulary, which constructs the penetration of flow in the interior and exterior spaces. The exterior stepped space is a void in the wall, and it is also a space of psychological communion with nature, landscape, and the wind.

The interior exhibition consists of two parts. The first space is underground space; it is the actual operating purification space where the audience can actually see the process of water purification through the machines' working. And this exhibit can educate the importance of water and new awareness of environment. This is the essential space of living purification exhibition which commune with education, exhibition, and experience.

As water flows down, the upper exhibition space is integrated with the 1st and 2nd floor with various formats for new experience of water. The architects drastically suggested the creation of water environment inside and experiencing it. In the void space that penetrates the 1st and 2nd level, water experience starts with the variously changing water dropping down like rain from the sky. And the audience can experience the nature playing with dropped water at the pond on the 1st floor. Also they communicate with the water through interactive media, and the experience of walking in the fog can be realized in the smog glass box. All exhibition area provides serial experiences in order to feel varied sensibilities with five senses. Adding education, exhibition, library, and resting spaces on the 2nd floor, spare space is created. And rooftop space becomes opened landscape.

YoonGyoo Jang + UnSangDong Architects

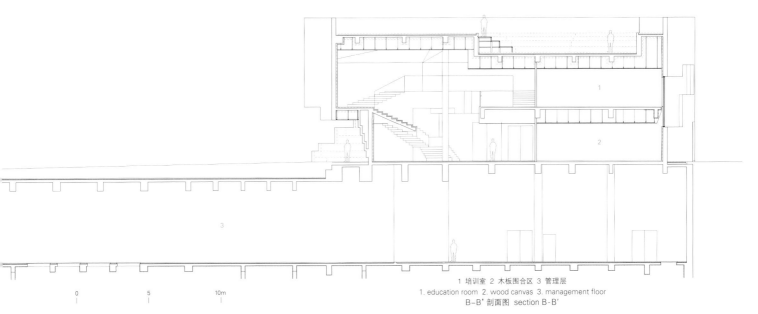

1 培训室 2 木板围合区 3 管理层
1. education room 2. wood canvas 3. management floor
B-B' 剖面图 section B-B'

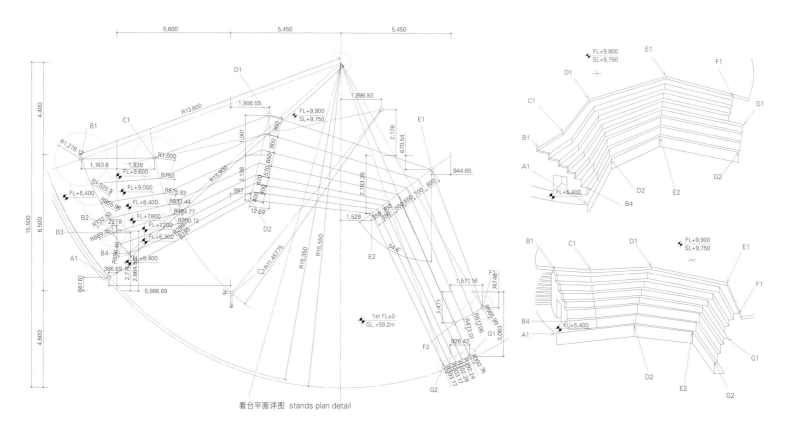

看台平面详图 stands plan detail

1. exposed concrete/stain (white)
2. floor reinforcing agent
3. THK150 plain concrete/
   machine gypsuming
   (#8@150x150 wire mech)
4. rubber-modified aspha waterproof/
   mortar
5. water-borne paint
6. THK12.5 gypsum board 2 ply
7. THK85 heat insulating board
8. light weight steel frame (C-STUD)
9. concrete
10. light weight steel frame (M-bar)
11. THK9.5 gypsum board 2 ply
12. putty/water borne paint
13. T12 laminate floor or T2 natural floor
14. hot-water heating
15. T40 lightweight aerated concrete

C-C' 剖面图 section C-C'

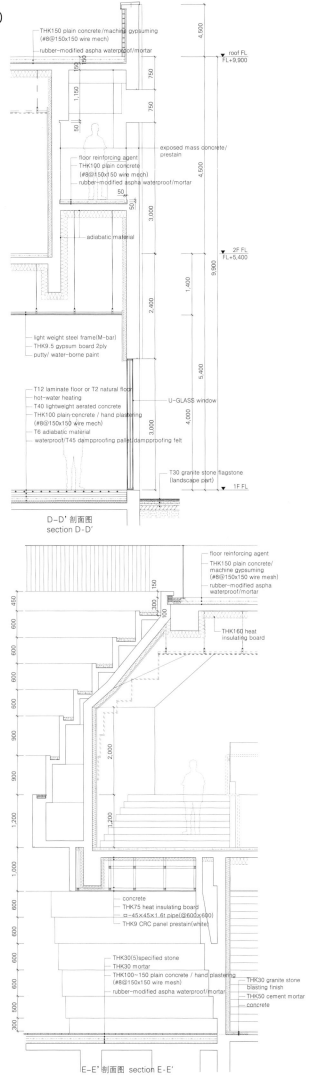

D-D' 剖面图
section D-D'

E-E' 剖面图 section E-E'

文化街
——松洞文化与福利中心

城市再生将会使被废弃的和被污染的环境焕然一新,城市需要这样的城市再生理念。当前城市再生理念在新城镇发展战略中得到了很好的应用,意在为现在千疮百孔的城市提供新项目和有形的建筑,来修复和振兴城市的环境、工业、经济和文化。这样做不是进行简单的建筑审美控制,而是需要构建新的交通流线系统,来给环境带来新的转变。

作为公共广场的建筑

本建筑设计摆脱了政府办公室的设计理念,独树一帜,营造出一种距离感。通过精心设计不同元素、考虑不同兴趣和引入包含文化的令人印象深刻的项目,松洞文化与福利中心成为宽敞的三维广场,同时也成为容纳不断变化发展的文化和福利的开放式建筑,是市民最喜欢的地方,也是市民享受福利和利益的主要场所。这是一座开放的公共建筑,是市民举行活动、开展文化和信息交流的枢轴。

体现绿色和城市结构的建筑

本建筑设计通过灵活运用水平的街道和广场,创造了水平、垂直和三维的广场空间,将城市结构与建筑整合为一体。建筑师通过添加不同元素、考虑公众不同兴趣和引入令人印象深刻的项目,创建了一个活力无限的文化和福利中心。另外,可移动的循环装置、运动内容和文化内容都明显体现了新的公民意识。再者,设计中也考虑到了未来的和实验性的变化。在这个城市结构中,不仅包括了建筑物所构建的结构,还包括城市中不可缺少的绿色元素。文化街成为人们丰富生活、稳定情感的地方。

Culture Street
SeongDong Cultural & Welfare Center

The city requires the concept of urban regeneration which renews the abandoned and polluted environment. The current concept of urban regeneration is well used as the strategy of new town development. It has the meaning of creating rehabilitation and revival of urban environment, industry, economy, and the culture by providing new programs and physical architectural environments to the decaying existing cities. Rather than simple architectural esthetic controls, the composition of new circulation system, which induces new transformation of environment, is required.

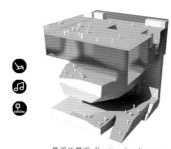
悬浮的景观 floating landscape

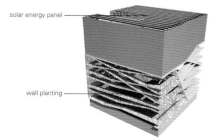
能量表皮 energy skin

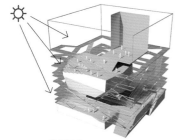
可持续性项目 sustainable program

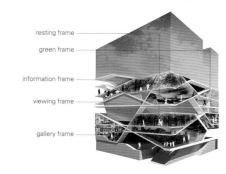

原始项目 original program

变化的项目 shaking program

生态&画廊 eco & gallery

项目合并 program consolidation

框架 frame

项目重新布局 program reshuffle

Architecture as a Public Plaza

The architecture, breaking away from the concept of government office, which creates exclusiveness and distances, is suggested. A Cultural & Welfare Center is suggested as an open three-dimensional plaza by planning different elements, interests, and impressive programs that contain culture. It becomes an open architectural place which accommodates changing culture and welfare at the same time. It became the citizenry's favorite place, and it is also the main place for welfare and its benefit. And it is the open public architecture. A pivot of civil activity, cultural, and informative events, is suggested.

Architecture as a Set of Green and Urban Structure

The urban structure provides a horizontal, vertical, and three-dimensional plaza in the building by actively accommodating horizontal streets and plaza. And the architecture, which integrates the urban structure, is suggested. A living welfare & culture center is proposed by adding diverse elements, interests, and impressive programs. Also moving circulation devices, movement contents, and cultural contents actively accommodate new senses of citizenry. And futuristic and experimental changes have been suggested. Not only the structures constructed by buildings, but also the green elements which are necessary in the city have been included in the urban structure. And it functions as a device for approaching the richness and emotional stability.

YoonGyoo Jang + UnSangDong Architects

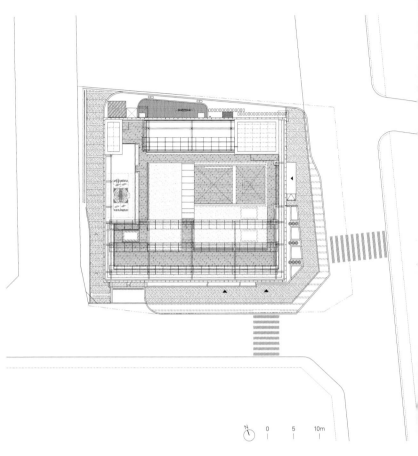

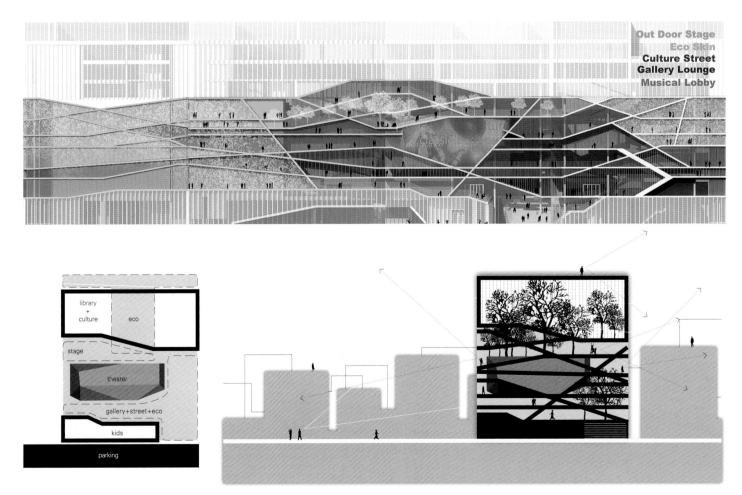

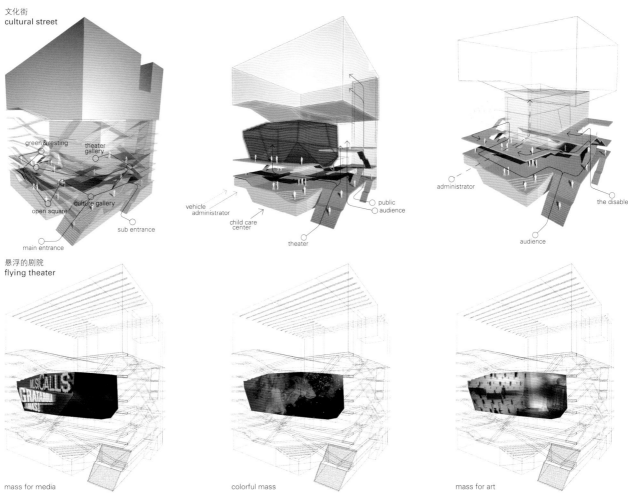

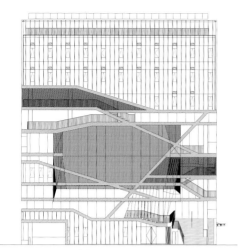
南立面 south elevation

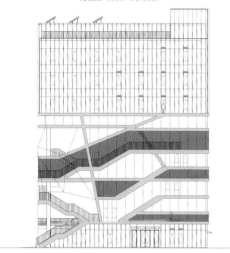
东立面 east elevation

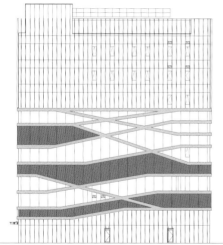
北立面 north elevation

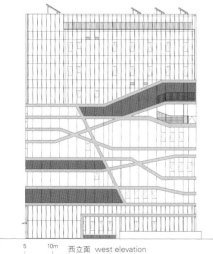
西立面 west elevation

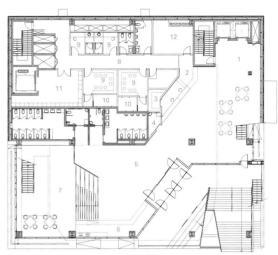

1 剧院大厅 2 售票亭 3 仓库 4 哺乳室 5 剧院主大厅 6 咖啡亭 7 休息区 8 走廊 9 更衣室 10 淋浴室 11 工具存储室 12 管理办公室 1. theater hall 2. ticket booth 3. stockroom 4. lactation room 5. theater main lobby 6. cafe booth 7. rest zone 8. corridor 9. dressing room 10. shower bath 11. toolmaker's storage 12. management office
二层 second floor

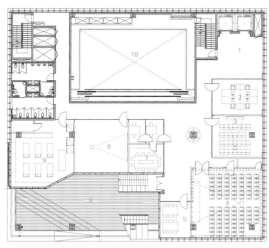

1 妇女福利中心大厅 2 管理办公室 3 讲座 4 多功能礼堂 5 女士休息室 6 女士花园休息室 7 咨询办公室（小组）8 程序室 9 烹饪教室 10 屋顶塔楼
1. women's welfare center lobby 2. management office 3. lecture room 4. multipurpose auditorium 5. women lounge 6. women garden lounge 7. counseling office (group) 8. program room 9. cooking classroom 10. fly tower
五层 fifth floor

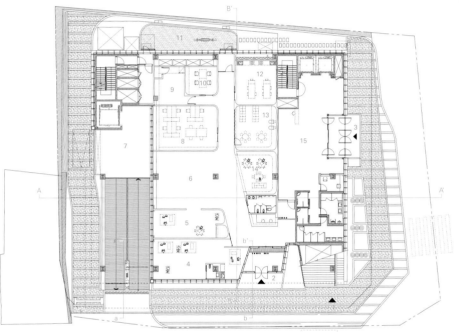

1 主入口 2 儿童广场入口 3 次入口 4 婴儿护理设施 5 儿童信息中心 6 游戏室 7 卸货平台 8 办公室 9 书本/教学用视听材料室 10 咨询室 11 儿童游乐场 12 程序室 13 资料室 14 儿童图书馆 15 大厅
1. main entrance 2. children plaza entrance 3. sub entrance 4. infant caring facilities 5. childcare information center 6. romper room 7. unloading dock 8. office 9. text book/teachware room 10. counseling office 11. children's playground 12. program room 13. reference room 14. children library 15. hall
一层 first floor

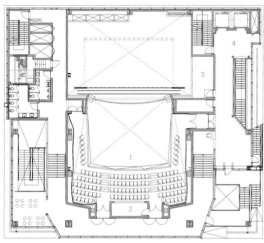

1 礼堂 2 控制亭 3 控制室 4 剧院大厅 5 升降机机房
1. auditorium 2. control booth 3. control room 4. theater hall 5. lift machine room
四层 fourth floor

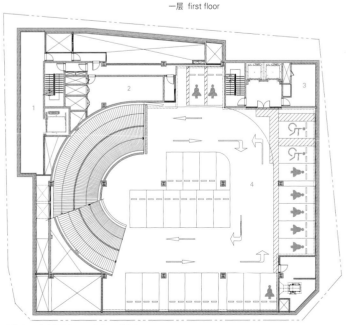

1 存储室 2 中央控制室 3 风扇房 4 停车场 1. storage 2. central monitor room 3. fan room 4. parking lot
地下一层 first floor below ground

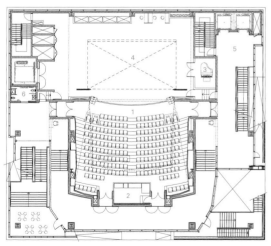

1 礼堂 2 控制亭 3 钢琴存储室 4 剧院舞台 5 剧院大厅 6 卫生间
1. auditorium 2. control booth 3. piano storage 4. theater stage 5. theater hall 6. toilet
三层 third floor

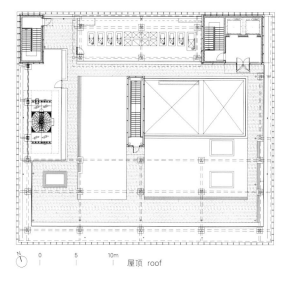

屋顶 roof

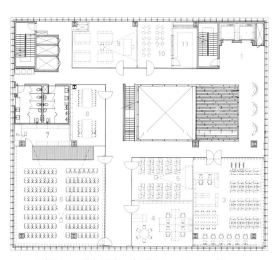

1 图书馆大厅 2 休息室平台 3 电子信息室 4 办公室 5 计算机教室
6 视听室 7 准备室 8 讨论室 9 讲室 10 小吃店 11 厨房
1. library hall 2. lounge deck 3. digital information room 4. office
5. computer lecture room 6. audio-visual room 7. preparation room
8. discussion room 9. lecture room 10. snack bar 11. kitchen

七层 seventh floor

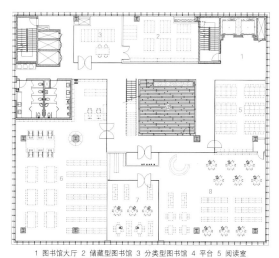

1 图书馆大厅 2 储藏型图书馆 3 分类型图书馆 4 平台 5 阅读室
6 文学馆 7 家庭阅读室 8 儿童阅读室
1. library hall 2. preservation library 3. arrangement library 4. deck 5. reading room
6. literature room 7. family reading room 8. children reading room

六层 sixth floor

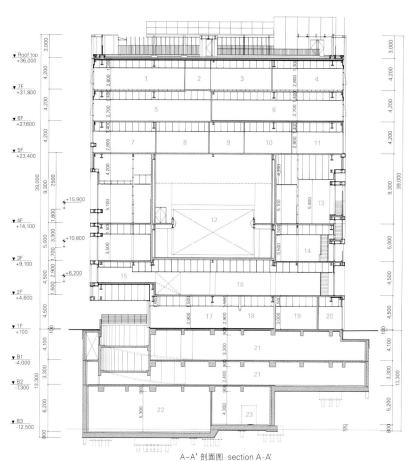

A-A' 剖面图 section A-A'

1 视听室
2 走廊
3 计算机讲室
4 电子信息室
5 文学馆
6 儿童阅读室
7 烹饪教室
8 准备室
9 咨询办公室
10 走廊大厅
11 讲室
12 礼堂
13 E.V大厅
14 剧院大厅
15 休息区
16 剧院主大厅
17 游戏室
18 儿童图书馆
19 大厅
20 管理办公室
21 停车场
22 机械室
23 电气室
24 办公室
25 小吃店
26 平台
27 储藏型图书馆
28 女士花园休息室
29 剧院舞台
30 哺乳室
31 仓库
32 剧院管理办公室
33 婴儿护理设施
34 儿童护理信息台
35 资料室
36 程序室
37 空气处理单元室
38 存放雨水的水槽
39 存放可回收雨水的水槽
40 化粪池

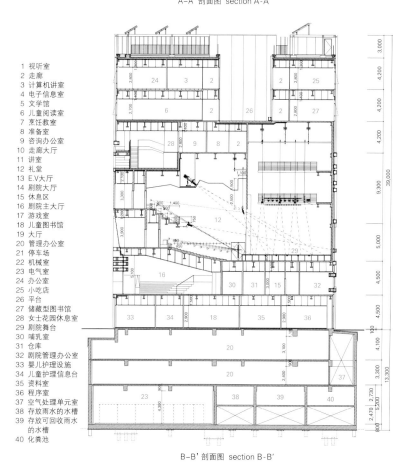

B-B' 剖面图 section B-B'

1. audiovisual room 2. corridor 3. computer lecture room 4. digital information room 5. literature room
6. children reading room 7. cooking classroom 8. preperation room 9. counseling office 10. corridor hall
11. lecture room 12. auditorium 13. E.V hall 14. theater hall 15. rest zone 16. theater main lobby
17. romper room 18. children library 19. hall 20. management office 21. parking lot 22. machine room
23. electric room 24. office 25. snack bar 26. deck 27. preservation library 28. women garden lounge
29. theater stage 30. lactation room 31. stock room 32. theater management office 33. infant caring facilities
34. child care information 35. reference room 36. program room 37. air handling unit room 38. rainwater tank
39. recycling rainwater tank 40. septic tank

项目名称：SeongDong Cultural & Welfare Center
地点：656-323,SeongSu-dong, SeongDong-gu, Seoul, Korea
建筑师：YoonGyoo Jang, ChangHoon Shin, SungMin Kim, MeeYoung Lee
项目团队：SamYeol Ryu, MinTae Kim, HyeLim Seo, HyeJoon Ahn, WonIl Kim, BooYoung Ahn, MiJung Kim, EunChong Jo
结构工程师：Thekujo
电气和机械工程师：HIMEC
甲方：Municipality of SeongDong-gu
用地面积：2,204m² 总建筑面积：1,014.69m² 有效楼层面积：9,558.75m²
结构：steel framed reinforcement concrete
设计时间：2009.6—2010.1 施工时间：2010.1—2012.9
摄影师：©JongOh Kim - p.153, p.156~157, p.160[bottom]
©Sergio Pirrone - p.155, p.160[top], p.161, p.164, p.165

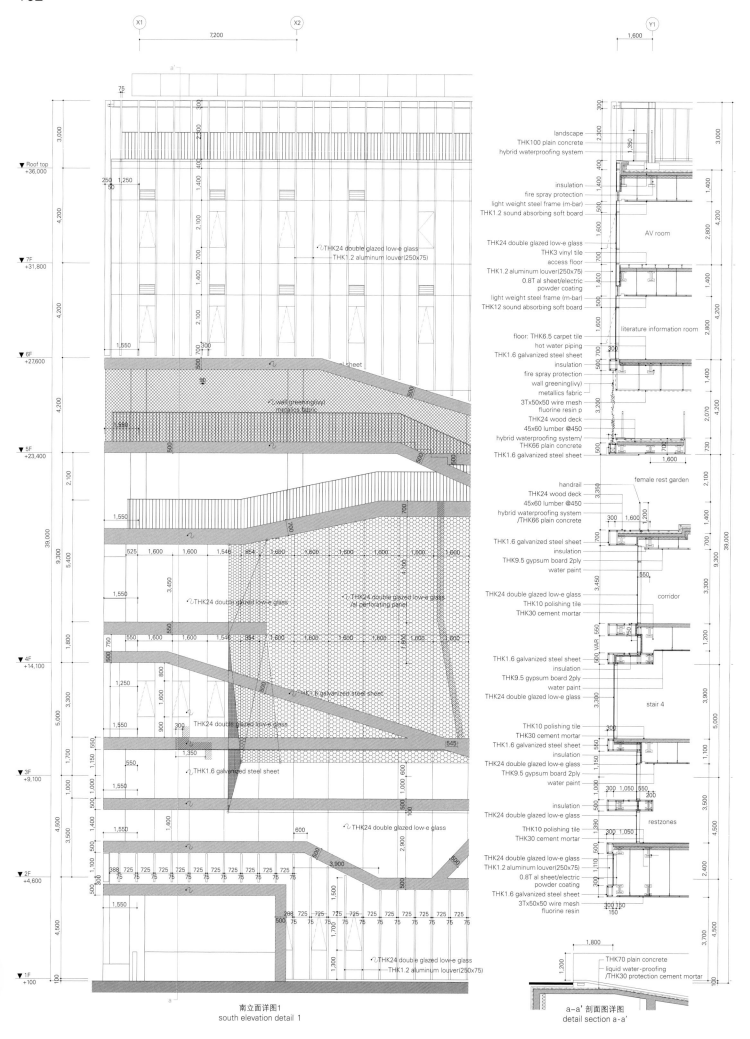

南立面详图1 / south elevation detail 1

a-a' 剖面图详图 / detail section a-a'

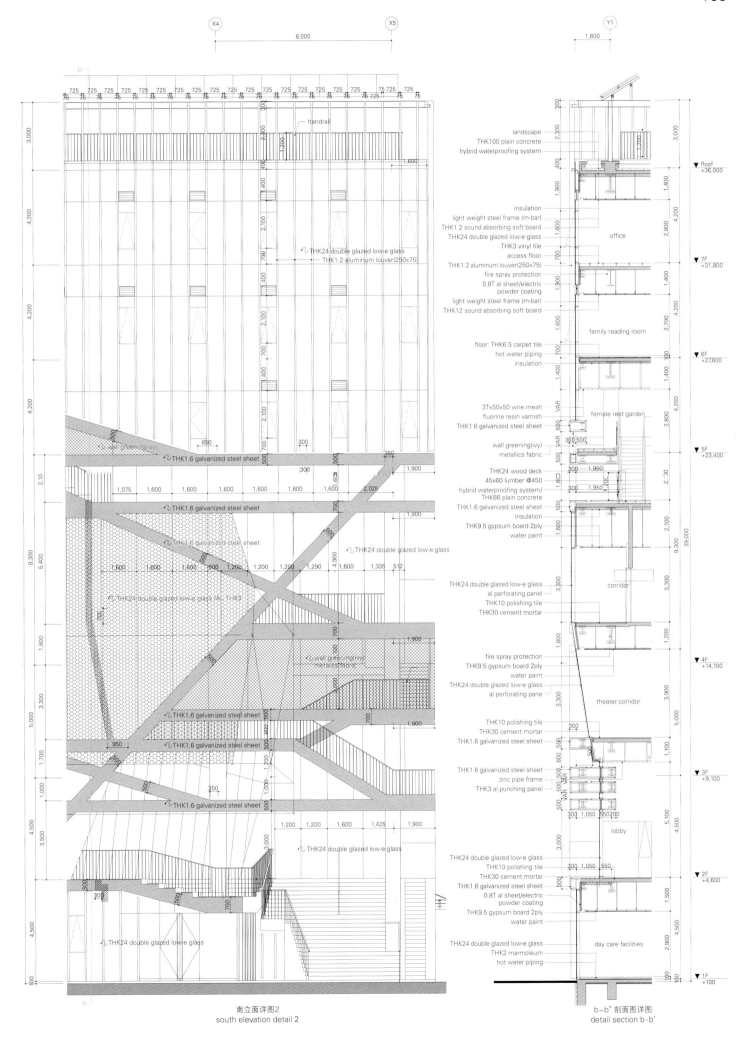

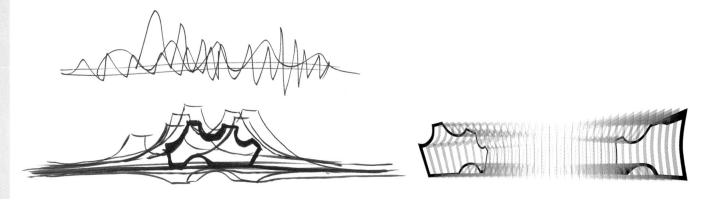

运动想象
——2012韩国YeoSu世博会现代汽车展馆

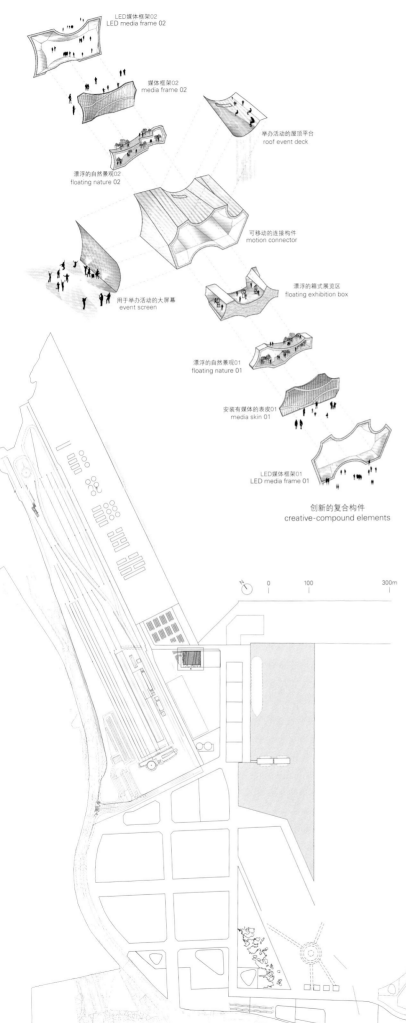

本设计希望通过2012韩国YeoSu世博会现代汽车展馆,来改变现代汽车集团的企业品牌形象,即在人类历史中创造全新的交通方式。

展馆的理念是"运动想象"。在这一理念的指导下,展馆建筑将人们对新的未来潮流的想象与现代汽车集团的运动融合为一体,再现了前所未有的尽情想象未来的可能性。展馆设计表达了现代汽车集团的精神,并通过再现自然与城市以及能源的再生所带来的无限可能性,来表现了技术的可实现性。

建筑所要展现的"运动"主要通过动态的、全新的剖面形式来实现。象征着三维的"蓝色海浪"运动的各种波浪形剖面相互重叠,连为一体,与不断变化的企业品牌形象的建筑表达相联系。

入口广场没有采用海浪形剖面,取而代之的是动感十足的、简单的盒子形剖面,面朝广阔的大海。

形成反差对比的高度连为一体,在富有雕塑感的建筑上实现了动态的运动。另外,将海洋波浪的运动与入口处运动的画面相连,象征着现代汽车集团拥有多种多样的精神。建筑起着连接的作用,连接着相互冲突的企业形象,如理性与感性、艺术与技术、过去与未来、变化与积累、简朴与多样。

建筑师希望设计实现建筑的动量,即实现蓝色海洋形象这一构造。这一建筑动力象征着现代汽车集团向世界发出的声音,这一剖面中蕴含的各种各样的运动和能量、高山的形状以及波浪都富有了象征意义。三维海浪这一建筑语言与多岛海公园的优美景色和谐地融为一体。在创造巨大的共振和回声的同时,还展现了品牌形象传遍城市每一个角落的场景。建筑师把展馆设计成一个大"音箱",吸收着大自然的能量,同时向这座海滨城市展现了企业形象。尤其是在夜晚,室内的媒体墙使整栋建筑外形变得清晰,这座媒体墙主要是利用锥形横截面来反射光线,播放媒体信息。

三维海浪空间展现了由自然和城市的各种形象和空间组成的复杂城市。

零重力区由现代汽车和钢铁行业的展览空间,城市的街道和广场、森林、休息区和游戏区组成,这处充满想象力的空间涉及到人类、环境、城市、技术和能源的问题,它们都与建筑息息相关。

像中庭内部的一些空展区,可以通过立方体式矩阵箱对空间加以改变。设计从建造一个白色立方体开始,白色立方体是一个标准单元和所有构件的基础。它们构成墙体,用机械操作来进行搬运,非常引人注目。这个室内的概念、景观墙,代表了改变运动的哲学。

Motion Imagination
2012 YeoSu EXPO Hyundai Motor Group Pavilion

Through 2012 YeoSu EXPO Hyundai Motor Group Pavilion, architectural work is suggested to convert the corporate brand image, which is creating the innovation of transportation in human history.

This architecture is combined with the imagination of new future waves and the motion of Hyundai Motor Group though the concept, "Motion Imagination". This is reproducing possibility of amplifying the futuristic imagination that is never achieved until now. It expresses their spirit and the possibility of technology by reproducing infinite possibility of nature and urban, as well as the energy.

The architectural issue of "motion" is proposed through the new formation of dynamic section. The section of various waves, signifies three-dimensional "Blue Ocean Wave" movement which is overlapped and connected. This also connects with the architectural expression of constantly changing corporate brand image.

One section at the entry square has dynamic displacement of wave, and the simple box shaped section outlook over the sea. Contrast elevations are connected so that dynamic motion is realized to sculptural architecture. Also, connecting the ocean side and the entry side frames of movements is the symbolic composition that the enterprise embraces various spirits. It is the architecture of link. The link connects the conflicting corporate images such as logical reason and emotion, art and technology, past and future, changes and accumulation, and simplicity and diversity.

The architects intend the momentum of architecture, configuration of blue ocean image, to be realized. It is the configuration of wave and vibration of Hyundai Motor Group to the whole world. The various movements and energy of the section, shape of mountain, and the wave of natural water are turned into figurative image. Architectural language of three-dimensional waves is to be harmonized with the landscape of DaDoHae Park. While creating huge resonance and echo, the brand scenario spreading to the city is embodied. The architecture is proposed as a large "sound box", which absorbs the energy of nature and releases the corporate image towards the maritime city. Especially at night, the media wall of interior creates luminous formation. It releases light and media while reflecting with single cross-sectional tapered shape.

In the space of three-dimensional waves, the complex urban is composed with various images and spaces of nature and cities. Zero gravity like space are composed with Hyundai Motor and steel sector's exhibition spaces, street of the city and plaza, forest, resting area and playing space. The container of imagination embodying future issues such as human, environments, urban, technologies, and energy is architecturalized.

The empty exhibition spaces, for example the atrium inside, suggest the space of alteration through cubical matrix box. By starting with white cube which is a standard unit and basis of all matters, they compose the wall and are transited spectacularly by mechanical operation. This interior concept, spectacle wall, stand for the philosophy of transforming movement.

YoonGyoo Jang + UnSangDong Architects

项目名称：2012 YeoSu EXPO Hyundai Motor Group Pavilion
地点：Site in 2012 YeoSu Expo, Korea
建筑师：YoonGyoo Jang, ChangHoon Shin, SungMin Kim, MeeYoung Lee
项目团队：SangHeon Hyun, MinTae Kim, YoungDong Goh, SeungHyun Kang, CholMin Jang, MiJung Kim, MinSun Son, EunJin Koh, HyeSoo Kim
结构工程师：Thekujo 机械和电气工程师：HIMEC
室内设计师：USD, GL Associates
甲方：Innocean Worldwide
用地面积：1,960m² 总建筑面积：1,397.50m²
有效楼层面积：2,334.81m²
结构：steel frame construction
设计时间：2011.1—11 施工时间：2011.12—2012.4
摄影师：©Sergio Pirrone

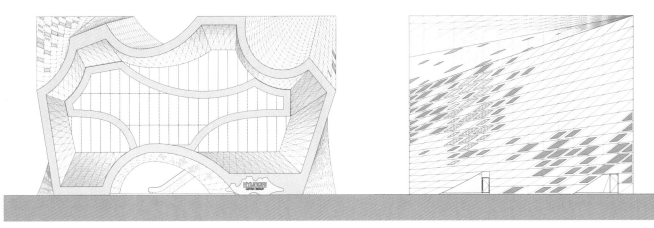

南立面 south elevation　　　　　　　　　　　　　东立面 east elevation

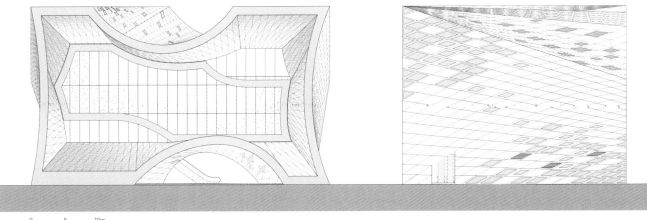

0　5　10m　　北立面 north elevation　　　　　　　　　　　　　西立面 west elevation

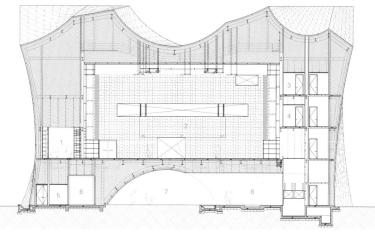

1 媒体走廊 2 展览空间 3 投影设备间 4 投影间 5 音响设备间 6 展厅 7 彩虹展览谷 8 入口展览室
1. media corridor 2. exhibition space 3. projection equipment room 4. projection room
5. sound equipment room 6. exhibition 7. rainbow exhibition valley 8. access exhibition
A-A' 剖面图 section A-A'

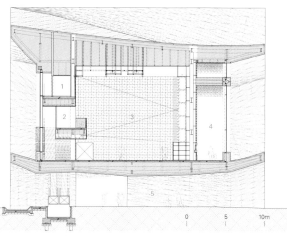

1 卫生间 2 展台 3 展览空间 4 媒体走廊 5 彩虹展览谷
1. toilet 2. exhibition stand 3. exhibition space
4. media corridor 5. rainbow exhibition valley
B-B' 剖面图 section B-B'

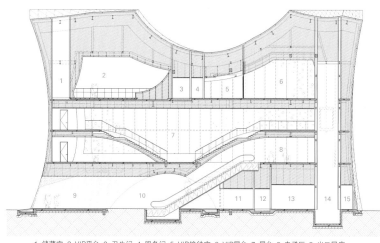

1 储藏室 2 VIP平台 3 卫生间 4 服务间 5 VIP接待室 6 VIP展台 7 展台 8 电梯厅 9 出口展廊
10 彩虹展览谷 11 员工操作休息室 12 VIP卫生间 13 VIP休息室 14 E.V正厅 15 E.V大厅
1. storage 2. VIP deck 3. toilet 4. service room 5. VIP reception room 6. VIP exhibition stand
7. exhibition stand 8. escalator hall 9. exit exhibition corridor 10. rainbow exhibition valley
11. staff operating lounge 12. VIP toilet 13. VIP lounge 14. E.V pit 15. E.V hall
C-C' 剖面图 section C-C'

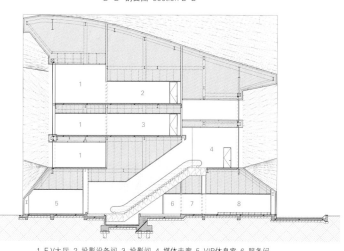

1 E.V大厅 2 投影设备间 3 投影间 4 媒体走廊 5 VIP休息室 6 服务间
7 员工操作休息室 8 指导休息室
1. E.V hall 2. projection equipment room 3. projection room 4. media corridor
5. VIP lounge 6. service room 7. operating staff hall 8. guide lounge
D-D' 剖面图 section D-D'

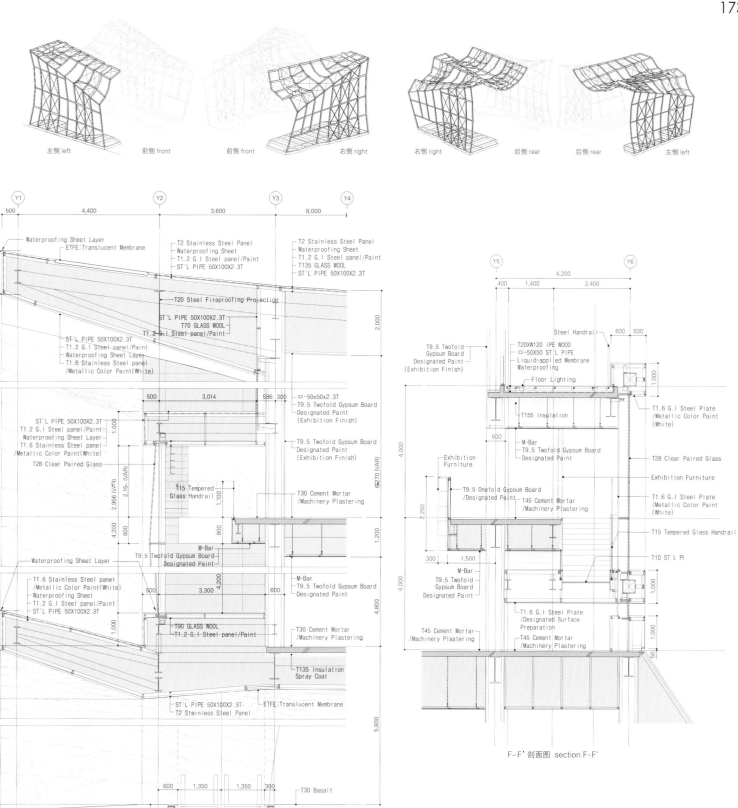

E-E' 剖面图 section E-E'

F-F' 剖面图 section F-F'

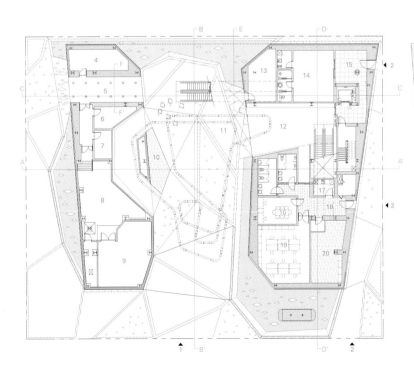

1 主入口 2 VIP入口 3 员工入口 4 储藏室 5 出口展览走廊 6 音响设备室 7 防火间 8 机械室 9 电气室
10 举办活动的舞台 11 彩虹展览谷 12 入口展览室 13 员工操作休息室 14 VIP休息室 15 VIP门厅
16 小组办公室 17 员工操作服务间 18 员工操作大厅 19 操作办公室 20 指导休息室

1. main entrance 2. VIP entrance 3. staff entrance 4. storage 5. exit exhibition corridor
6. sound equipment room 7. fire prevention room 8. machinery room 9. electric room 10. event stage
11. rainbow exhibition valley 12. access exhibition 13. staff operating lounge 14. VIP lounge 15. VIP foyer
16. group office 17. staff operating service room 18. staff operating hall 19. operating office 20. guide lounge

一层 first floor

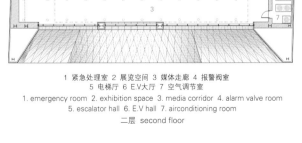

1 紧急处理室 2 展览空间 3 媒体走廊 4 报警阀室
5 电梯厅 6 E.V大厅 7 空气调节室

1. emergency room 2. exhibition space 3. media corridor 4. alarm valve room
5. escalator hall 6. E.V hall 7. airconditioning room

二层 second floor

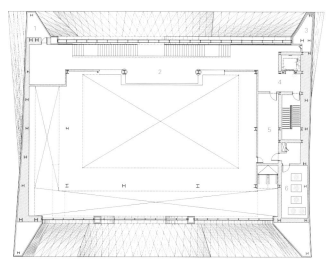

1 紧急处理室 2 展览空间 3 报警阀室 4 E.V大厅
5 投影室 6 空气调节室
1. emergency room 2. exhibition space 3. alarm valve room
4. E.V hall 5. projection room 6. airconditioning room
二层（分层） stratified second floor

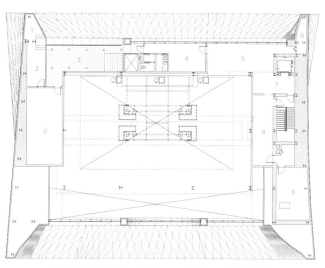

1 紧急处理室 2 储存室 3 VIP平台 4 VIP接待室 5 VIP展台 6 报警阀室
7 E.V大厅 8 投影设备间 9 空气调节室
1. emergency room 2. storage 3. VIP deck 4. VIP reception room
5. VIP exhibition stand 6. alarm valve room 7. E.V hall
8. projection equipment room 9. airconditioning room
三层 third floor

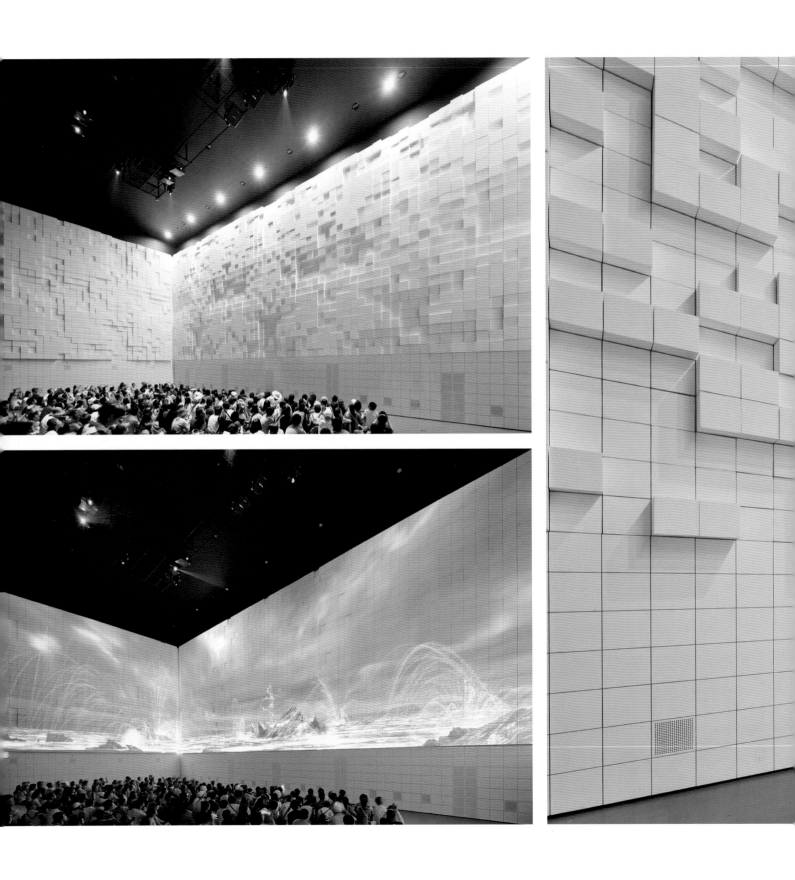

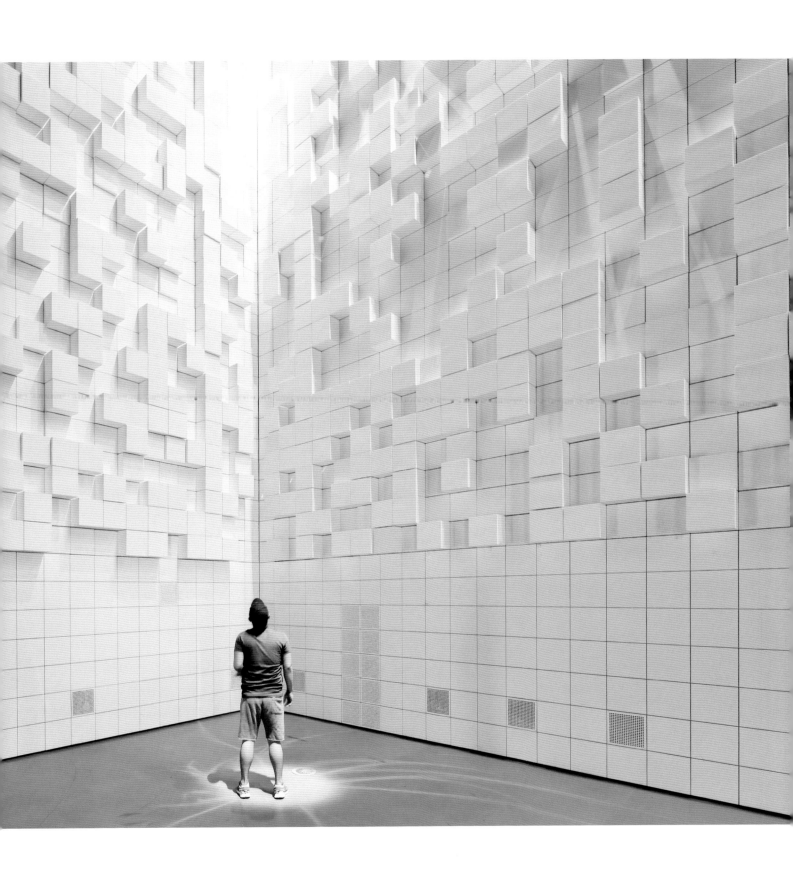

Rooftecture
—— 能源与绿色住宅

全球变暖、自然资源的枯竭以及能量干涸等，这些由消耗型社会所引起的威胁正日益严重，环境问题成为人们无法避免的重要问题。特别是大城市中无休止的扩张与消耗，如人口、建筑与其他设施等，都急需新型城市住宅来进行解救。这里阐述了生态系统的需求，而这些需求是从自然与人工建筑的结合和共生中产生的。

能源节约绿色住宅

作为住宅的主题之一，建筑师建议能源节约绿色住宅要优于零能耗。这个利于环境的系统通过将"景观式建筑"与"生态式建筑"结合起来，使其本身在自然与人类生活中共存。它包含了各种绿色因素，如结构系统、材料、空间构成和景观，它们能够丰富人类的生活。同时建筑师还将绿色科技应用在可持续住宅的优化系统中。因为这些住宅满足了日常生活中的自然因素。因此，它在建筑与自然之间和谐共存，成为自然友好型住宅。

Rooftecture住宅：住宅设计与绿色科技相结合

建筑师的目标不仅仅是将绿色科技与毫无设计头脑的建筑外形相结合，而是致力于将技术与建筑精妙地结合在一起。

而此时此刻，这种环境友好型住宅在完成所有技术要求的基础上得以竣工。建筑师采用Rooftecture能源与绿色住宅，由能源系统与景观构成。在这座住宅中，屋顶，这一最重要的发挥遮盖的作用部分，采用了抽象的地面形状，并且将技术的建筑表皮与能源系统结合起来。屋顶斜坡的坡度能够使其能源消耗达到最少，但是却能够获得最多的太阳能。同时它还设有能够利用水能的曲形外表以及室外露台，这些元素结合在一起，形成了独一无二的屋顶。这座建筑带有适当的皱折、斜坡和拱肋，屋顶形状，如同一座连续的山体，有效地利用了自然资源，如太阳、水、土地以及风。从理智方面和感情方面来说，这是一座外形及其系统均与自然相关联的建筑。新型有机建筑从自然界中走出，由科技赋予灵感，而这种技术是从所有的自然生物中获取指引原则。

Rooftecture
Energy + Green Home

Global warming, exhaustion of natural resources, energy draining, etc., as these dangers caused by consumption-oriented society are on the rise, environmental issues become important subject People cannot avoid. Especially, endless expansion and consumption in metropolitans, regarding population, architecture, infra, etc., face to enquire new type of urban housing. Here lies the demands for ecological system, which can be acquired from the combination and symbiosis of nature and artificiality.

Energy Plus House

As one theme of housing types, the architects suggest Energy Plus Green Home surpassing energy zero. This environmental friendly system gains symbiosis between nature and human life through the combination of "Landscape Architecture" and "Ecological Architecture". It includes diverse green factors, which make the human life enriched such as structural system, material, spatial composition and landscape. Green technologies are combined in optimum system for sustainable housing. As residents meet the natural factors in daily life, it becomes nature-friendly house harmonized in architecture and nature.

Rooftecture Dwelling; Housing Design Integrated with Green Technology

The architects' aim is not only to juxtapose green technologies with just a mindless architectural shape but to made stringent effort to combine technology and architecture in a subtle way.

At this point of the time, environmental-friendly dwelling is completed according to technical needs. They adopt "Rooftecture", compound of energy system and landscape. Here, the roof, the most important part considering a role of shelter, adopts the shape of land in the abstract and combines technological skin with energy generating system. Minimum energy loss, the degree of roof slope for gaining maximum solar energy, curved forms to utilize water energy, outdoor terrace... the combination of these elements composites on the sole roof of Rooftecture with appropriate creases, slopes and ribs. The roof, like a consequent mountain shape efficiently uses natural resources such as the sun, water, earth and wind. It is a rational and emotional housing taking shape and system from nature. New organic Architecture is extracted from nature inspired technology, which studies the principle of all natural creatures. YoonGyoo Jang+UnSangDong Architects

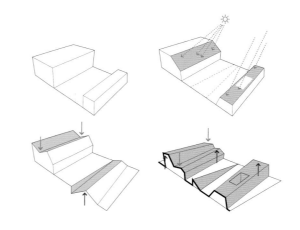

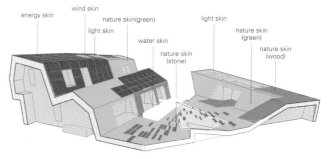

太阳能 solar energy

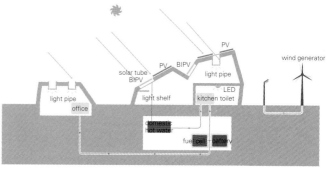

地热 geothermal

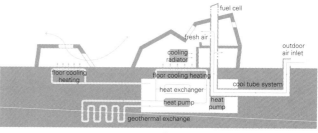

水 water

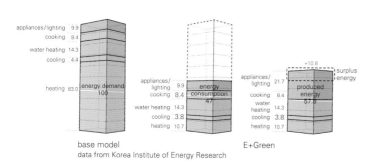

data from Korea Institute of Energy Research

能源绿色住宅是为4人家庭而设计的。
假设一个家庭每年的能源要求基数为100, 那么建筑师就能为其节省53% (采用被动的方式)。
然后, 建筑师却能生产57.8%的能量, 因此, 该住宅反而能增加4.8%的能量。

Energy Green Home is designed for 4 people family.
It is assumed that annual energy demand for the family is 100 and the architects save 53% in a passive way. Furthermore, they produce energy of 57.8%. Consequently they gain 4.8% of surplus energy.

项目名称：Rooftecture _ Energy + Green Home
地点：JeonDae-ri PoGok-eub CheoIn-gu GyengGi-do, Korea
建筑师：YoonGyoo Jang, ChangHoon Shin, YounSoo Kim, MeeYoung Lee
项目团队：YoungEun Choi, HyeJoon Ahn, HoJin Kim, HyeLim Seo, MiJung Kim, JiHye Kim
结构工程师：Thekujo
机械与电气工程师：HIMEC
室内设计：USD
甲方：Kolon Global Corporation
用地面积：5,525m² 总建筑面积：957.40m² 有效楼层面积：1,837.04m²
结构：reinforced concrete
设计时间：2009.5—2010.6 施工时间：2010.6—2011.4
摄影师：©Sergio Pirrone

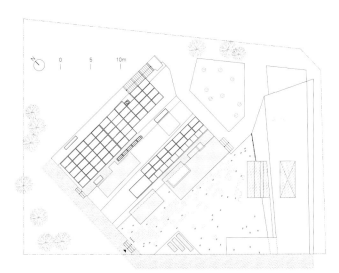

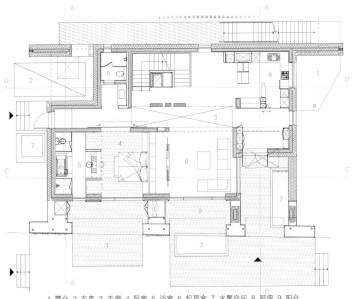

1 露台 2 车库 3 走廊 4 卧室 5 浴室 6 起居室 7 水景空间 8 厨房 9 阳台
1. terrace 2. garage 3. corridor 4. bedroom 5. bathroom 6. living room 7. water space 8. kitchen 9. balcony
一层 first floor

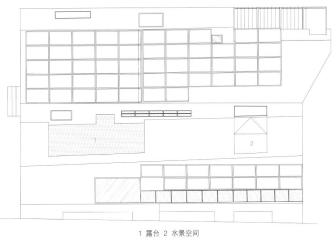

1 露台 2 水景空间
1. terrace 2. water space
屋顶 roof

1 机械间 2 电气室 3 系统间
1. mechanical room 2. electric room 3. system room
地下一层 first floor below ground

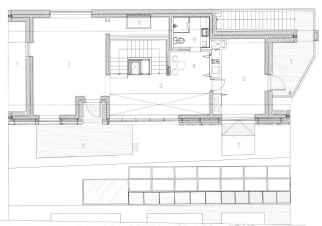

1 露台 2 画廊 3 走廊 4 家庭活动室 5 卧室 6 浴室 7 水景空间
1. terrace 2. gallery 3. corridor 4. family room 5. bedroom 6. bathroom 7. water space
二层 second floor

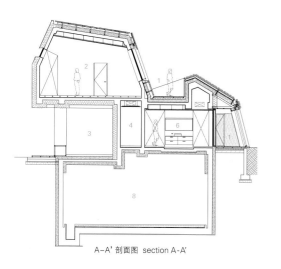

A-A' 剖面图 section A-A'

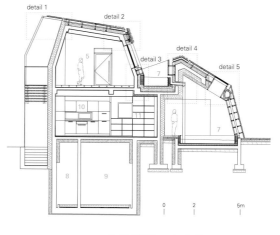

B-B' 剖面图 section B-B'

1 露台	1. terrace
2 画廊	2. gallery
3 车库	3. garage
4 走廊	4. corridor
5 卧室	5. bedroom
6 浴室	6. bathroom
7 水景空间	7. water space
8 机械室	8. mechanical room
9 系统间	9. system room
10 厨房和餐厅	10. kitchen & dining room
11 入口	11. entrance
12 楼梯间	12. stair hall
13 家庭活动室	13. family room
14 起居室	14. living room

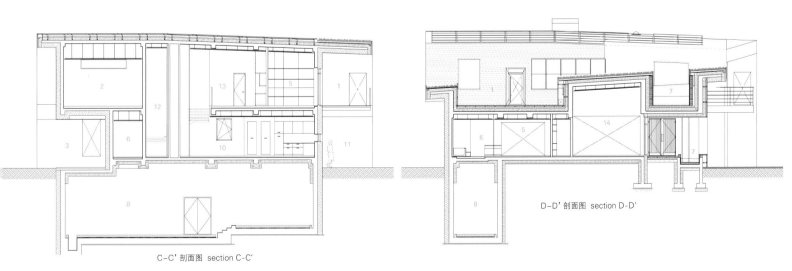

C-C' 剖面图 section C-C'

D-D' 剖面图 section D-D'

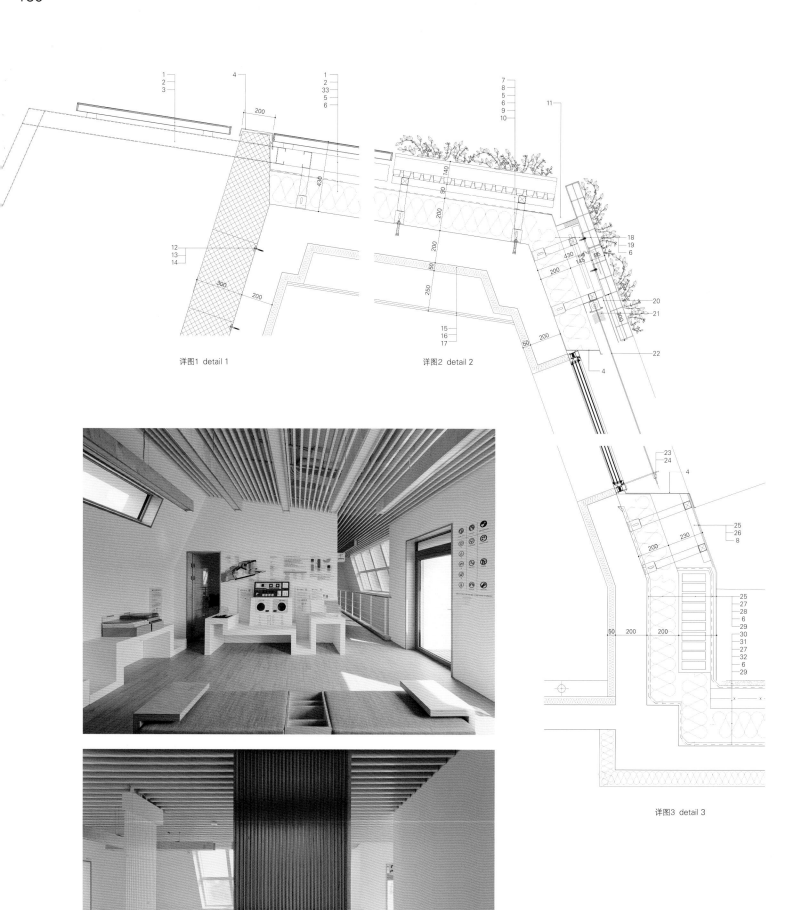

详图1 detail 1

详图2 detail 2

详图3 detail 3

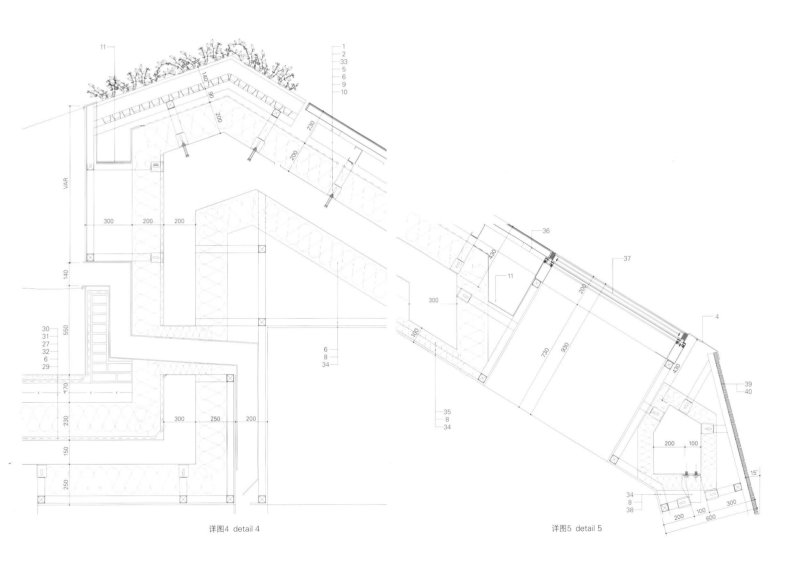

详图4 detail 4   详图5 detail 5

1. 980x1,620x50 PV panel
2. STL pipe 30x60x2. 1T
3. AL bar-50x120
4. THK3 AL flashing
5. waterproof coating
6. THK200 wet insulator
7. green roof system
8. STL pipe 45x45x2. 1T
9. STL bracket L=50x60x80x5T
10. STL anchor bolt 3/8xL=100
11. rainwater gutter
12. THK3 cement free fiber mortar
13. THK300 expandable polystyrene
14. PVC tracka
15. THK50 insulator
16. THK9 gypsum board
17. THK9 specified finish
18. green panel
19. STL pipe 45x75x2. 1T
20. cover panel
21. venetian blade
22. wire
23. bottom
24. tension bracket
25. THK specified porcelain tile
26. THK12 waterproof plywood
27. liquid waterproof twice
28. cement brick laying
29. urethane waterproof
30. rainwater tank
31. THK30 landscape black gravel
32. THK100 plain concrete
33. C-STL channel 100x50x20x2. 3T
34. THK12 CRC panel
35. THK100 wet insulator
36. PV panel
37. BIPV
38. STL pipe 100x100x4. 5T
39. THK20 wood deck
40. THK2 vibration panel

>>84
Cruz y Ortiz Arquitectos
Was founded by Antonio Cruz[left] and Antonio Ortiz[right]. Both were born in Seville, Spain and graduated from ETSAM. They are members of the Andalusian Architects' Association. Their offices are located in Seville and Amsterdam, and they handle many international architectural projects. Won the international competitions in 2012 such as the competition for University Campus in Saclay, Paris, Stadium in Zurich, Stadium in Lugano, Switzerland.

>>70
Arditti + RDT Arquitectos
Mauricio Arditti[middle] graduated from the National Autonomous University of Mexico and his trajectory integrates over 50 years of professional experience. At the midpoint of his career, he received a second generation of architects incorporating his two sons Arturo Arditti[right] and Jorge Arditti[left], both graduated from Anahuac University in Mexico City. They all have complemented specialized studies abroad, in schools like Harvard University and the Massachusetts Institute of Art, and have combined their knowledge and experi-

>>40

Morphosis Architects
Is an interdisciplinary practice involved in rigorous design and research that yields innovative, iconic buildings and urban environments, founded in 1972. With founder Thom Mayne serving as design director, the firm today consists of a group of more than 40 professionals, who remain committed to the practice of architecture as a collaborative enterprise. With projects worldwide, the firm's work ranges in scale from residential, institutional, and civic buildings to large urban planning projects. Over the past 30 years, Morphosis has received 25 Progressive Architecture awards, over 100 American Institute of Architects (AIA) awards, and numerous other honors.

>>26

SANAA
Kazuyo Sejima formed a team with Ryue Nishizawa in 1995. Since then, she has produced many projects jointly with him. Was born in Ibaraki Prefecture in Japan, 1956 and established Kazuyo Sejima & Associates in 1987. Was a visiting professor at Ecole Polytechnique Federale de Lausanne and Princeton University. Ryue Nishizawa was born in 1966 and graduated from Yokohama National University with a master's degree in architecture. In 1977, he established the Office of Ryue Nishizawa. Currently, he is a visiting professor at Harvard Graduate School of Design.

>>12

Herzog & de Meuron
Was established in Basel, 1978. Has been operating by senior partners; Christine Binswanger, Ascan Mergenthaler and Stefan Marbach, with founding partners Pierre de Meuron and Jacques Herzog from the left. Ascan Mergenthaler joined the firm in 1998 and became a partner in 2004. Established Herzog & de Meuron's US office in 2001 and led a number of high profile projects in the US. Since 2009, he has been senior a partner in charge of international projects in Asia, North and South America and Europe, among which are extension of the Tate Modern and its surrounding areas, London, UK.

YoungBum Reigh
Is currently a professor at the Graduate School of Architecture of KyongGi University. Studied architecture at Seoul National University and received his Ph. D. from Graduate School of Architectural Association School of Architecture in London. Has actively been involved in Urban Action Network, a Non-governmental Organization in the field of community organization and community design. Based on the experience, he has published in a series of books on the issues of creative city and urban regeneration.

Marta González Anton
Is an architect, works and lives in Rotterdam, the Netherlands. Studied in Spain and Italy. Before moving to Rotterdam she collaborated with several local offices in Spain and worked on publication activities. Is currently carrying out a project research about the contemporary design process between the Netherlands and Spain.

Silvio Carta
Is an architect and critic based in Rotterdam. Lives and works in the Netherlands, Spain and Italy where he regularly writes reviews and critical essays about architecture and landscape for a diverse group of architecture magazines, newspapers and other media. In 2009, he founded the Critical Agency™ | Europe.

>>98
Gonçalo Byrne Arquitectos
Gonçalo Byrne is the founder and senior CEO of Gonçalo Byrne Arquitectos. For the last 35 years his works have been internationally recognized. Was a professor in Coimbra, Lausanne, Venice, Mendrisio, Leuven, and Harvard.

>>128
YoonGyoo Jang + UnSangDong Architects
UnSangDong Architects Cooperation experiments and realizes "Conceptual Architecture" which tries to represent many functions of an architect as cultural contents. Their goal is to realize the depth of an architect by letting UnSangDong Architects, Gallery JungMiSo, UnSangDong Publication, and UnSangDong Art communicate each other. The cooperation won AR Award given by Architectural Review, in 2007 and also got Vanguard Award given to an innovational architect by Architectural Record in 2006. In 2001, they were selected one of world's forty architects by Japanese journal 10+1. Their projects include "Kring Cultural Complex Center", "Gallery Ye", "Life & Power Press", "Hi Seoul Festival" and "Ocean Imagination", etc. Through all of those projects, they are trying to show an architectural works that cross over both an architect and art.

>>116

AF6 Arquitectos

Has been operating by Miguel Hernández Valencia, Esther López Martín, Juliane Potter, Francisco José Domínguez Saborido, Ángel González Aguilar from the left. Recently, they won the competition for Triana Ceramic Museum in Seville(2009), Apartment building in Seville(2010), Intervention in common spaces of social housing in Seville (2012).

>>56

Groupo Arquidecture

is a team led by Ricardo Combaluzier left, William Ramírez right and Josefina Rivas middle. Was founded in 2005 under the name of 4A Arquitectos, time after with the accumulated experience in the architectural area and considering the needs of modern society, evolves into Groupo Arquidecture. Their best quality in the field of design and development of architecture and urban projects has been proved by receiving many major awards. The recent project, Gran Museo del Mundo Maya, won Emblematic Building Ibero American CIDI Award 2012, and Best Pathfinder Project at the Partnerships Awards 2012.

C3, Issue 2013.6

All Rights Reserved. Authorized translation from the Korean-English language edition published by C3 Publishing Co., Seoul.

© 2013大连理工大学出版社
著作权合同登记06-2013年第214号

版权所有·侵权必究

图书在版编目(CIP)数据

博物馆的变迁 / 韩国C3出版公社编；于风军等译. —大连：大连理工大学出版社，2013.9
（C3建筑立场系列丛书；30）
ISBN 978-7-5611-8226-0

Ⅰ．①博⋯ Ⅱ．①韩⋯ ②于⋯ Ⅲ．①博物馆—建筑设计 Ⅳ．①TU242.5

中国版本图书馆CIP数据核字(2013)第216719号

出版发行：大连理工大学出版社
　　　　　（地址：大连市软件园路80号　邮编：116023）
印　　刷：北京雅昌彩色印刷有限公司
幅面尺寸：225mm×300mm
印　　张：12
出版时间：2013年9月第1版
印刷时间：2013年9月第1次印刷
出 版 人：金英伟
统　　筹：房　磊
责任编辑：张昕焱
封面设计：王志峰
责任校对：高　文

书　　号：ISBN 978-7-5611-8226-0
定　　价：228.00元

发　行：0411-84708842
传　真：0411-84701466
E-mail：12282980@qq.com
URL：http://www.dutp.cn